Essentials of Flowcharting

Third Edition

Essentials of Flowcharting

Third Edition

Michel H. Boillot
Gary M. Gleason
L. Wayne Horn

Pensacola Junior College

ωcb
Wm. C. Brown
Company Publishers
Dubuque, Iowa

Copyright © 1975, 1979, 1982 by Wm. C. Brown Company Publishers

Library of Congress Catalog Card Number: 81–68490

ISBN 0–697–08151–6

Third Printing, 1983   2–08151–02

Printed in the United States of America

Contents

Preface

This text presents a logical approach to programming via flowcharting. It is suitable for use in courses which emphasize any of the popular high level languages. It provides a systematic development of programming techniques which include: formula evaluation, loop control, counting and accumulating, end of file detection and control breaks. Additional topics such as file processing, arrays, structured programming and subprograms provide opportunities for application of the basic programming techniques.

Chapter 1 contains a brief overview of the relationship of flowcharting to problem solving and computers. In Chapter 2 the standard flowcharting symbols are defined and illustrated in numerous examples. The concepts of loops and counting are developed in Chapter 3. Chapter 4 contains an introduction to file processing techniques, a brief look at system flowcharts, and a short section on structured programming. In Chapter 5 the concepts of arrays and subscripts are developed and the subprogram concept is introduced.

Appendixes provide illustrative programs in BASIC, COBOL and FORTRAN. An introduction to the BASIC language and to decision tables are also included.

Having used the first and second edition as a supplement in an introductory level computer science course, this third edition has given the authors an opportunity to improve significantly upon the previous editions. Suggestions from students and instructors have led to a presentation of material in which topics are presented in a more logical and "teachable" sequence. Many exercises have been rewritten and new exercises have been added to illustrate real life application problems.

The third edition reflects the decreased importance of punched cards in data processing. As interactive computing has become a prevalent mode of operation in most educational computing centers/labs, references to the punched cards as an input medium have been deleted in the text and in the exercises. The appendix on decision tables is new in this edition. The authors have found that reference to decision tables can be very helpful for some students who have difficulty with networks of decisions, such as those required for many of the exercises in Chapters 2 and 3. The authors welcome communications from any user.

M. Boillot, Ed.D.
L. W. Horn, Ed.D.
G. M. Gleason, C.P.A.
Pensacola Junior College
Pensacola, Fl. 32504

1 Problems, Procedures and Computers

PROBLEMS AND PROCEDURES 1-1

The digital computer has become an indispensable aid in problem solving in almost every area of human endeavor. From law to medicine, engineering to library science, business administration to elementary education, statistics to literary research, the computer is used as an extension of man's inquiring mind. This alliance of man with the computer's processing capabilities has and will continue to increase man's understanding of himself and his universe. Yet as a piece of machinery with no "mind" of its own other than the ability to carry out arithmetic and logical operations at high speed, the computer is ultimately dependent on the human mind to supply it with the necessary instructions to solve any type of problem. This set of man-generated instructions, consisting of the necessary steps required to derive a solution to a given problem, is called a *program*. A computer program is a specific example of a procedure. A procedure is a representation of a method for solving a problem.

Procedures are common phenomena in everyday life; a recipe for baking a cake, a set of instructions for building a model airplane, a set of maxims for living the "good life" are all examples of procedures which man might employ to achieve a desired goal. Good procedures in a computing environment share several important features:

1. Precision
2. Finiteness
3. Effectiveness
4. Input
5. Output

A procedure which meets these criteria is called an *algorithm*. In an algorithm the instructions used must be *precise*. A recipe which calls for 1/4 teaspoon of salt is more precise than one which calls for a pinch of salt, since a pinch implies different amounts to different people. An algorithm must be *finite;* it must terminate after execution of a finite number of steps. An algorithm must be *effective;* it must solve the problem for which it is proposed in an efficient manner. An algorithm calls for some *input;* that is, information or material to be processed according to the instructions of the algorithm to produce certain *output*. This output is the solution of the problem for which the algorithm was devised.

The process by which problems are solved may be visualized as in Figure 1.1. Essentially a problem-solving process can be described as a two distinct stage process. In the first stage a procedure is devised to solve the particular problem. This is the "mental processing" stage. The mind is the processing unit; the input consists of the problem to be solved; the output is the procedure. The second stage consists of the execution of that procedure by carrying out the instructions by some "mechanical processing" means, i.e., some machine, a computer, or man's physical motion. This implementation stage generally calls for some input to generate the desired output, i.e., the solution to the original problem.

1-2 PROCEDURES AND COMPUTERS

Inasmuch as a procedure represents the description of how a particular problem is to be solved, it may be formulated in numerous ways: pictorially, verbally, or in written form. The translation of a procedure into another form is a very delicate one as it involves the translation of a mental process into a set of directives. The "thinking man" may still be able to carry out a set of instructions that are incomplete or possibly confusing because he can exercise judgment. One cannot expect such behavior on the part of computers, as these are extremely limited in their "thinking ability." For that reason it is imperative that procedures exhibit the five distinctive features of an algorithm discussed in Section 1-1. It is also important to understand the computer's characteristics, functions, and mode of operation, since these will, in many respects, determine the manner in which procedures may be formulated in a computing environment.

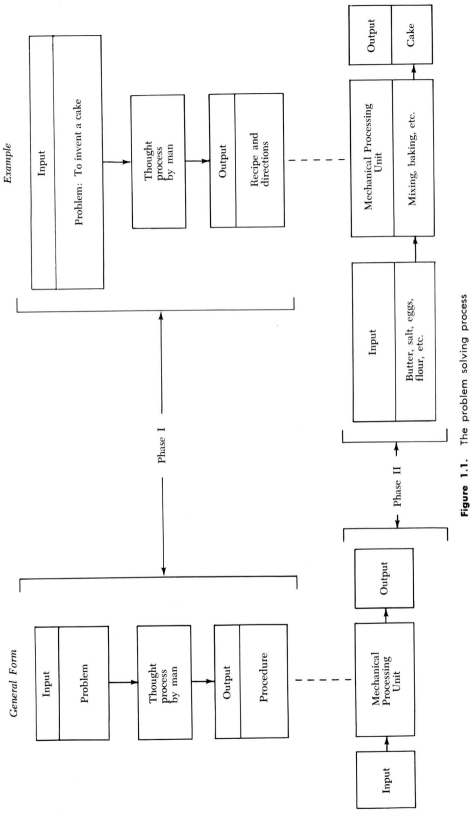

Figure 1.1. The problem solving process

A digital computer is composed of five basic components (see Figure 1.2) whose functions can be described as follows:

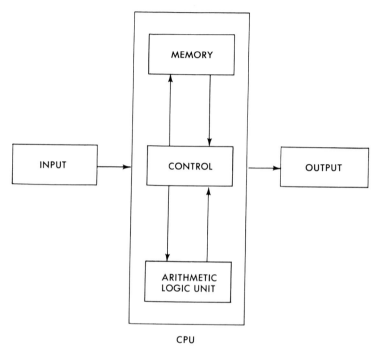

Figure 1.2.   Functional units of a computer

1. Input: Data is read into the memory of the computer from some external source.
2. Output: Information stored in the memory is written or displayed on some type of output medium.
3. Memory: The memory is used to store the set of instructions corresponding to a procedure, with additional information to be used by the computer.
4. Arithmetic Logic Unit: Performs arithmetic operations and logical operations such as comparing, and branching.
5. Control: Coordinates all components and ensures proper execution of instructions in memory.

Physical devices used to accomplish the above functions vary from one computer configuration to another. Common input devices are punched card readers and terminals; common output devices include printers, typewriters, and terminals. The combination of memory, control, and arithmetic logic unit is usually referred to as the *Central Processing Unit* (CPU).

The memory unit is always divided into locations which are addressable. We may think of the memory as a large group of pigeon holes, each with a name (*symbolic address*) by which we can refer to its contents. The amount of information that can be stored in each of these locations varies from one computer to another; for the purpose of this introductory discussion, we will assume that each location contains an instruction or a number.

Execution of a *program* (i.e., the set of instructions representing a particular procedure) can only take place when that program is in memory. The control unit then processes each instruction one at a time, until the desired result is obtained. Ordinarily the control unit processes instructions sequentially; sometimes though, as a result of some decision in the program, it may become necessary to bypass a set of instructions. This is called *branching*. The following example will help illustrate the above concepts.

A record has three numbers on it (see Figure 1.4.): a code, a number of hours worked, and a rate of pay. Let us write a set of instructions to compute the pay. If the code read is zero, a bonus of fifty dollars is to be added to the pay; in either case the results are to be written. A procedure for solving this problem might consist of the following steps:

1. Read the three numbers from the record and place them in locations called CODE, H, and R.
2. If the code is equal to 0, branch to step #5.
3. Compute the product of H and R, place the result in a location called PAY.
4. Branch to step #6.
5. Compute the product of H and R, add 50 to the product and place the result in a location called PAY.
6. Write the content of PAY onto the output device.
7. Stop.

The above steps could be formalized in a hypothetical programming language as follows:

1. READ CODE, H, R
2. IF CODE = 0 GO TO 5
3. PAY = H * R
4. GO TO 6
5. PAY = H * R + 50
6. WRITE PAY
7. STOP

Figure 1.3 shows how this program might appear when stored in memory prior to execution by the computer (before computer carries out the instructions).

① READ CODE, H, R	② IF CODE=0 GO TO 5	③ PAY = H * R	④ GO TO 6
⑤ PAY=H*R+50	⑥ WRITE PAY	⑦ STOP	
Ⓗ	Ⓡ	⟨CODE⟩	⟨PAY⟩

Figure 1.3. Pigeonhole memory device with sample program

Does this set of instructions meet the criteria established for a valid algorithm?

1. Input is introduced by the READ operation.
2. Output is produced when the value for PAY is written out.
3. The procedure is finite since it terminates in five or six instructions depending on whether a branch is taken in instruction #2.
4. Each instruction is precise.
5. It is effective since it solves the stated problem.

We can then conclude that this program represents an algorithm.

EXECUTION OF A PROGRAM 1-3

Assume that the preceding program is now stored in memory as shown in Figure 1.3 and that the control unit is about to execute the program. If the record shown in Figure 1.4 is presented to the computer, the first instruction of the program will cause the three values 1, 40, and 3 on the record to be stored respectively in the locations CODE, H, and R as shown in Figure 1.5.

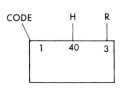

Figure 1.4. Sample record

①	②	③	④
READ CODE, H, R	IF CODE=0 GO TO 5	PAY = H * R	GO TO 6
⑤	⑥	⑦	
PAY=H*R+50	WRITE PAY	STOP	
Ⓗ	Ⓡ	(CODE)	(PAY)
40	3	1	

Figure 1.5. Memory after execution of first step

The second instruction causes comparison of the value stored in CODE in zero. Since the content of CODE is not equal to zero, the branch to the instruction in location five is not taken. Instead the instruction in location three is executed next. This instruction causes the product of H and R to be placed in PAY. The contents of memory after execution of this instruction are shown in Figure 1.6. The instruction in location four causes the next instruction to be taken from location six. The instruction in location six causes the content of location PAY to be written onto some output medium thus displaying the results from the program. The instruction in location seven then causes the system to wait for a new program or for some other action to be taken.

① READ CODE, H, R	② IF CODE=0 GO TO 5	③ PAY = H * R	④ GO TO 6
⑤ PAY=H*R+50	⑥ WRITE PAY	⑦ STOP	
Ⓗ 40	Ⓡ 3	(CODE) 1	(PAY) 120

Figure 1.6. Memory after execution of the third step

The actual form that the instructions may take vary from one computer to another. The intent of this book is not to delve into the details of the various programming languages but rather, illustrate a language in which algorithms may be expressed as a set of general instructions parallelling a computer's way of thinking. This language is called *flowcharting*. The remaining chapters of this book address themselves to the study of flowcharts and useful techniques for solving problems with computers.

EXERCISES 1-1

| 0 | 40 | 3 |

Figure 1.7. A second sample record

1. Suppose the record shown in Figure 1.7 is to be read by a computer as programmed in Figure 1.3. What output will be produced?
2. Consider the following procedure:

Basic White Bread

Bring to a boil in a small sauce pan:

    1/2 cup milk
    1 T. butter
    1 t. sugar
    1 t. salt

Add 1/2 cup cold water and let mixture cool to lukewarm. Meanwhile dissolve 1/2 cake yeast and 1 t. sugar in 1/4 cup warm water. When the milk mixture is cool, add the yeast mixture and beat in approximately 2 cups sifted flour. Knead in enough additional flour to form a dough that is moist but does not stick to the hands. Place in a greased bowl, cover with a damp cloth, and let rise until double in bulk, about 2 to 3 hours. Punch the dough down and knead briefly. Shape the dough and place in a greased loaf pan. Cover and let rise until double, about 2 hours. Place loaf in a cold oven, turn heat to 400°, and bake 15 minutes. Reduce heat to 350° and bake 20 minutes longer. Remove and allow to cool before slicing. Yield: 1 loaf.

Is this procedure an algorithm according to the criteria described in this chapter? Could an experienced cook execute the procedure?

3. Prepare a procedure similar to that shown in Section 1-2 to calculate the area of a triangle if the base and altitude are provided on a record.

4. Paychecks are to be prepared from time records showing the name, hours worked, and hourly wage for a group of employees. Prepare an algorithm for the controller who will write the checks.

2   Program Flowcharts

A *flowchart* is a diagram displaying the procedure used to solve a problem. It is a visual outline describing the instructions schematically. Inasmuch as a flowchart represents the implementation of a particular procedure to solve a problem, no specific familiarity with computers or computer languages is required. The organization and structure of a flowchart nevertheless should reflect a computer's "way of thinking," and to this end, a minimal knowledge of the general functions of a computer is necessary to help insure easy translation of the flowchart into a computer language.

Flowcharts are useful at all stages in the problem-solving process. Initially the problem solver may express the overall formulation of the procedure by means of a rudimentary flowchart. Subsequently this flowchart can be refined or altered as the problem solver focuses his attention on specific details. Verification of the effectiveness of the procedure can then be accomplished by tracing all possible alternatives displayed in the flowchart. The flowchart may be useful in communicating a procedure from the problem solver to the computer programmer. From a flowchart, the programmer can derive the necessary code for computer processing. Finally, the flowchart serves as documentation for the computer program. It may be easier to understand the logic of the program by referring to the flowchart rather than to the actual program.

2-2 FLOWCHARTING SYMBOLS

There are five basic symbols used to draw program flowcharts. Each corresponds to one of the basic functions of a computer. The symbols (also referred to as *blocks*) represent instructions; the symbols we shall use are shown in Figure 2.1. Students who wish to draw

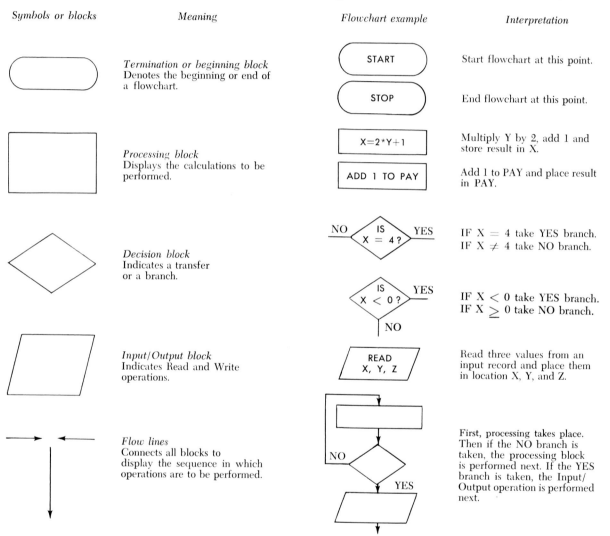

Symbols or blocks	Meaning	Flowchart example	Interpretation
	Termination or beginning block Denotes the beginning or end of a flowchart.	START	Start flowchart at this point.
		STOP	End flowchart at this point.
	Processing block Displays the calculations to be performed.	X=2*Y+1	Multiply Y by 2, add 1 and store result in X.
		ADD 1 TO PAY	Add 1 to PAY and place result in PAY.
	Decision block Indicates a transfer or a branch.	NO — IS X = 4 ? — YES	IF X = 4 take YES branch. IF X ≠ 4 take NO branch.
		IS X < 0 ? — YES / NO	IF X < 0 take YES branch. IF X ≥ 0 take NO branch.
	Input/Output block Indicates Read and Write operations.	READ X, Y, Z	Read three values from an input record and place them in location X, Y, and Z.
	Flow lines Connects all blocks to display the sequence in which operations are to be performed.	NO — ... — YES	First, processing takes place. Then if the NO branch is taken, the processing block is performed next. If the YES branch is taken, the Input/Output operation is performed next.

Figure 2.1.   Program flowcharting symbols

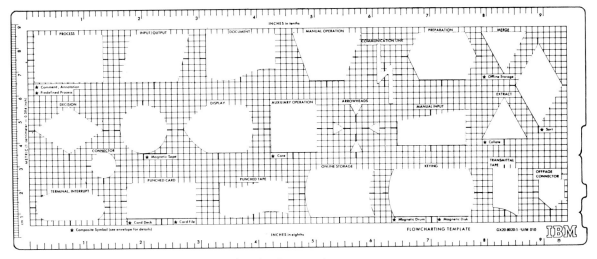

Figure 2.2. Flowcharting template

flowchart symbols precisely may want to use templates that are commercially available (see Figure 2.2).

The content of a block is called a *statement*. Thus, we may speak of an Input block but a Read statement. Consider, for example, the flowchart and sample input record shown in Figure 2.3.

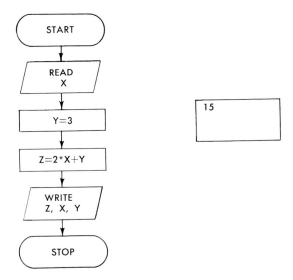

Figure 2.3. Flowchart and sample record

Figure 2.4. Contents of memory after second instruction

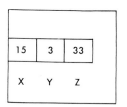

Figure 2.5. Contents of memory after third instruction

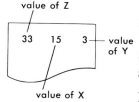

Figure 2.6. Sample output

The symbol ⟨START⟩ identifies the starting point in the

flowchart. The instruction /READ X/ will cause the computer to read a record. The value contained on that record (15 in this example) is stored in memory location X. By submitting records with different numbers, different values may be assigned to X. For this reason locations with symbolic names such as X are often referred to as *variables*.

The instruction ⟨Y = 3⟩ causes the value 3 (which is generated internally) to be placed in memory location Y. The contents of the two memory locations, X and Y, (variables) at this point are shown in Figure 2.4.

The instruction ⟨Z = 2 * X + Y⟩ causes the value of the expression on the right side of the equal sign to be evaluated and placed in the variable Z. In this case $2 * X + Y$ is evaluated as $2 * 15 + 3 = 33$. The value 33 is stored in Z. The contents of the three memory locations X, Y, Z now are shown in Figure 2.5.

The instruction /WRITE Z,X,Y/ will cause the contents of locations Z, X, and Y to be written onto some output device (perhaps a printer) as shown in Figure 2.6.

Finally the instruction ⟨END⟩ denotes the end of processing.

2-3 THE PROCESSING BLOCK

A rectangle is used whenever calculations or computations are to be performed. *Replacement statements* are used in the processing block to express the computations. The general form of a replacement statement and some examples of its use are shown in Figure 2.7.

A single variable must always be placed on the left hand side of the equal sign; it is the memory location into which the value of the expression is stored. Thus $3 * X + 2$ is *not* a replacement statement since no location is designed to store the result. The statement $3.1 = X$ is invalid since 3.1 is a constant; the left hand side of the equal sign

The general form of a replacement statement is

VARIABLE = EXPRESSION

where   VARIABLE is any symbolic name representing a memory location. (A variable generally starts with a letter of the alphabet.)

and   EXPRESSION is any combination of variables and/or constants linked by arithmetic operations. (Note that just one variable or a single constant may be classified as an expression. Parentheses may be used to indicate the order in which operations are to be performed.)

Examples of variables: PAY, HRS, X, PAY23
                      123, 4AB are invalid variables (start with digit)

Examples of expressions: 3.1, $-2.5°Y$, X^2+Y^2
                      (HRS − 40) ° RATE

Examples of replacement statements:

X = Y + 2   Evaluate Y + 2 and store the result in memory location X (call result X).

PAY = HRS * RATE + BONUS   Multiply the number in HRS by the number in RATE and add to it the number in BONUS. Place result in memory location PAY

Figure 2.7. The replacement statement and its uses

must be a variable and not a constant, besides how could you store the number in X in the number 3.1? However, the statement X = 3 is a valid way of assigning the value 3 to the variable X. All variables used in an expression must have been previously defined before the expression can be evaluated, otherwise, the results of a statement such as Z = 2 ° X + Y where X and Y have not been assigned a value previously, cannot be predicted.

The equal sign used in a replacement statement must be understood as a replacement sign rather than a mathematical equality.[1] Consider the statement X = X + 2. Algebraically there are no values of X

---

1. Many texts use the symbol ← to denote replacement to avoid the ambiguity of the two meanings of the equal sign.

satisfying this equality. Yet understood as a replacement statement, this statement is quite meaningful. If the value of X is 10 before the statement $X = X + 2$ is processed, the value of X after processing will be 12 since $X + 2 = 10 + 2 = 12$. The value of the expression replaces whatever value was contained in X previously.

Names for variables need not be restricted to one character as in algebra. Names can be formed to suggest the nature of the data under consideration. For example, the names PAY, HOURS, and RATE might be used in a statement such as PAY = HOURS * RATE. We will also adopt the convention that variable names may contain digits, but must start with a letter of the alphabet. For example, PAY1 and G231 are valid variable names.

The symbols commonly used in arithmetic expressions are as follows:

+	addition
−	subtraction
*	multiplication
/	division
**	exponentiation[2]

Note the use of the asterisk to indicate multiplication. This convention is necessary to avoid the ambiguities which may arise from other common ways of expressing multiplication. In some programming languages the symbol ** is used to denote exponentiation; that is, X^2 would be coded as X ** 2. In this text we shall, however, use the superscript for exponentiation.

Although the use of replacement statements is a common way of expressing computations, verbal description of operations may also be

used.[3] For example, the processing block $\boxed{\begin{array}{c} \text{ADD 1 TO} \\ \text{COUNT} \end{array}}$ might be

used instead of the block $\boxed{\text{COUNT} = \text{COUNT} + 1}$. In any case,

it is always possible to express a replacement statement by an equivalent verbal statement. Consider the following examples:

2. In some programming languages the symbol " ↑ " is used for exponentiation. For example, X^2 would be coded X↑ 2.

3. Replacement statements are used in scientific programming languages such as FORTRAN and BASIC, whereas verbal statements are generally used in business programming languages such as COBOL.

Verbal form	**Replacement statement form**
MOVE 10 TO TOTAL	TOTAL = 10
ADD PURCHASES TO INVENTORY	INVENTORY = INVENTORY + PURCHASES
SUBTRACT PAYMENT FROM AMOUNT DUE	AMOUNT DUE = AMOUNT DUE — PAYMENT
MULTIPLY HRS BY RATE GIVING PAY	PAY = HRS * RATE

The use of verbal descriptions is sometimes made to shorten otherwise lengthy replacement statements. Care must be exercised to write valid verbal statements, i.e., any such verbal statements must correspond to a replacement statement. For example $\boxed{\text{COMPUTE X} + \text{Y}}$ is invalid since there is no destination implied for the value computed. The block $\boxed{\begin{array}{c}\text{FIND THE}\\\text{TOTAL}\end{array}}$ is invalid since there is no value specified to place into TOTAL.

FIELDS, RECORDS, FILES 2-4

As noted in Chapter 1 computers can read and write data from a variety of devices or mediums: punched cards, magnetic tape, or terminals. Data on such mediums is organized by fields; for example, if punched cards are used, a *field* is defined as a group of related card columns. A field identifies a particular entity; a *record* is a group of related fields and as such describes many characteristics of the entity. A *file* is a collection of related records. In Figure 2.8 each record consists of four fields which describe or list the characteristics of a particular individual.

A group of related records constitutes a file, that is, a file consists of a set of records which all have identical formats. Figure 2.9 displays files consisting of three records each.

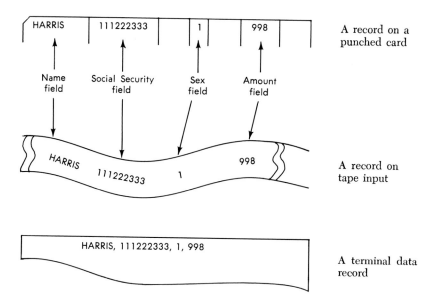

A record on a
punched card

A record on
tape input

A terminal data
record

Figure 2.8.   Records on a variety of mediums

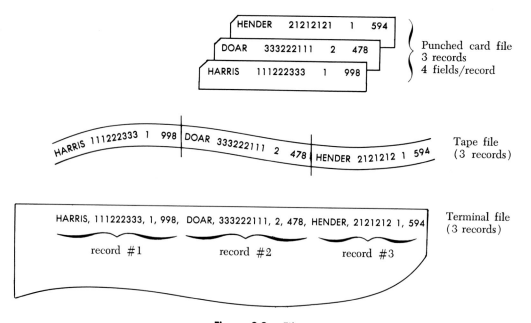

Punched card file
3 records
4 fields/record

Tape file
(3 records)

Terminal file
(3 records)

record #1   record #2   record #3

Figure 2.9.   Files

THE INPUT-OUTPUT BLOCK 2-5

The general form of the input-output block and examples of its use are shown in Figure 2.10. The command READ or WRITE is followed by a list of variables separated by commas. Each READ statement causes the reading of *one* record. For instance if a terminal is used, each input block causes a different group of fields to be read; if a line

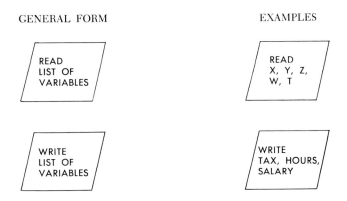

Figure 2.10. General form of input-output blocks

printer is used as an output device each WRITE statement causes one entire line to be printed out. A record can contain one or more values (fields). The number of variables in the input or output list determines the number of values that will be read from a record or written on a record.

The size of a record, i.e., the number of fields in a record is generally determined by the number of variables specified in the input/output list. For example, the block

READ NAME, SOCSEC SEX, AMT , with data record HARRIS, 111222333, 1, 998

will cause the characters HARRIS to be associated with the variable NAME, the values 111222333, 1 and 998 to be associated with the variables SOCSEC, SEX and AMT. That is, HARRIS will be stored in memory location NAME, 111222333 will be stored in memory location SOCSEC etc. Note that in this case the list of variables consists of four variables and hence the number of fields per record is four. Details as to how the group of characters (fields) on a record are separated from one another are specific to programming languages.

The command WRITE, followed by a list of variables, will cause the value of the specified variables to be written onto some output medium (paper form, tape etc.) by some output device (printer, terminal, etc.). For example with the values of the variables NAME, SOCSEC, SEX, AMT defined as above, the block:

The output line in this case is the output record. As with input, the way in which fields are separated on the output depends on the specific programming language used.

EXERCISES 2-1

1. Differentiate between a constant, a variable, an expression and a statement.
2. In what ways may a variable be assigned a value? i.e., in what different ways can we store a number in a memory location?
3. Are the following replacement statements valid?

   a. $13 = X$ e. $PAY = RATE * HOURS$
   b. $F = F$ f. $OVERTIME + REGULARTIME$
   c. $X + Y = 2 * F - 3$ g. $3.4 = 3.4$
   d. $I = I + 1$ h. $SUM = SUM + AMT$

4. Are the following valid processing blocks?

   a. | ADD PAYMENT |
      | TO TOTAL |

   d. | CALCULATE |
      | RATE |

   b. | SUBTRACT |
      | AMOUNT |

   e. | DIVIDE TOTAL |
      | BY COUNT GIVING |
      | AVERAGE |

   c. | MOVE AMOUNT |
      | TO AMOUNTOUT |

   f. | SUBTRACT CHARGE |
      | FROM NET |

5. Write processing blocks containing replacement statements equiv-
    alent to the valid blocks in exercise 4.
6. How does the flowchart in Figure 2.12 differ from that in Figure
    2.3? Does Figure 2.12 represent an algorithm?
7. Can you predict what values will be written by the flowchart in
    Figure 2.11? Is it an algorithm?

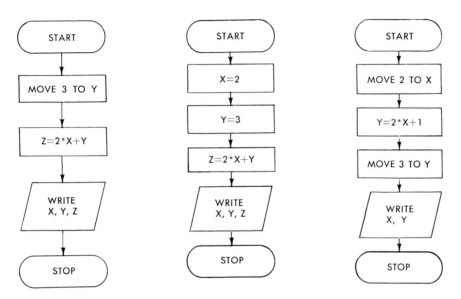

Figure 2.11.   **Figure 2.12.**   **Figure 2.13.**

8. What output will be produced by the flowchart in Figure 2.13?
9. What output will be produced by the flowchart in Figures 2.14,
    2.15, and 2.16 if the given data records are processed?
10. Draw flowcharts for each of the following sets of instructions:
    a. Read amount due and amount paid from a record. Subtract
        amount paid from amount due giving balance forward. Write
        out the balance forward for the account.
    b. Read a value for PAY and write it out.
        Set BONUS equal to 100.
        Compute PAY + BONUS and write the result.
    c. Read unit selling price and number of units sold from a record.
        Calculate and print the revenue generated.

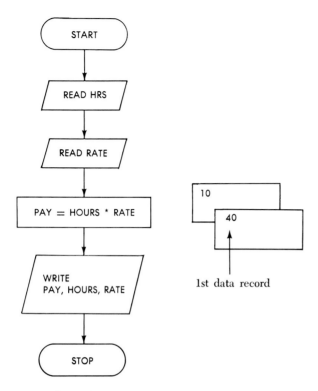

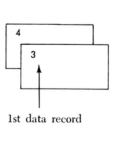

Figure 2.14.

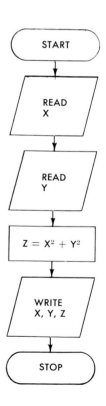

Figure 2.15.

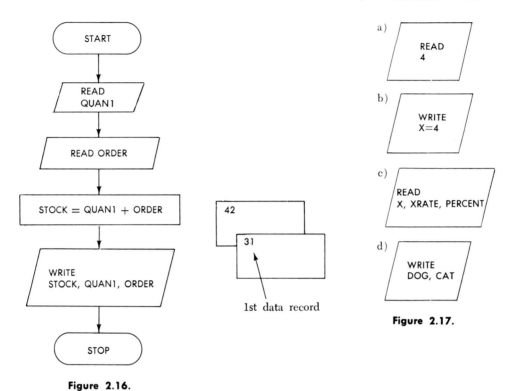

Figure 2.17.

Figure 2.16.

11. Which of the input/output blocks in Figure 2.17 are valid?
12. Determine the values of the variables after execution of the following input blocks and data records:

a) READ A, B, C   with data record   12,   11.6,   −13

b) READ HRS, RATE   with data record   12,   11.6,   14

c) READ A, B, A   with tape record   11   0   3

d) READ GRADE, GRADE, GRADE   with data record   12,   11.5,   −3

13. Draw a flowchart to read two numbers from one record and print the two numbers, their sum and their difference on one line.

14. Draw a flowchart to read three fields from one record: a principal P, an interest rate R and a time duration T. Compute and print the simple interest due. (Note: $I = P°R°T$)

15. Draw a flowchart to read a temperature given in centigrade and compute the equivalent Fahrenheit value by using the formula: $F = 9/5 ° C + 32$. Print the result.

16. Draw a flowchart to read the two sides A and B of a right triangle and compute and print the length of its hypothenuse, H. (Note: $H = \sqrt{A^2 + B^2}$)

17. Draw a flowchart to approximate the Julian date equivalent to the calendar date given in the form: month, day. The Julian date is the day of the year. January 1 has Julian date 1, February 2 has Julian date 33, December 31 has Julian date 365, etc. A formula to approximate the Julian date is $(month - 1) ° 30 + day$. Read from a record a month and a day and print its Julian date equivalent.

18. A record consists of three fields: a name, a rate of pay, a number of hours worked. Draw a flowchart to read one record and print.
    a—The name and pay on the same line.
    b—The name, rate of pay, number of hours, and pay on same line.
    c—The name on one line, the rate of pay on the second line and the pay on the third line.

19. Draw a flowchart to compute the area and circumference of a circle with the radius R read from a data record. The area of a circle is $\pi ° R^2$; the circumference of a circle is $2 ° \pi ° R$. ($\pi = 3.1416$)

20. The following fields are provided on a data record: name, gross-pay, federal income tax, withholding rate, miscellaneous deductions. FICA tax is deducted at the rate of 6.7% of gross pay. Draw a flowchart to compute and print net pay.

21. Mary Smith wishes to deposit $1000 in a savings account earning 12% for one year. She has a choice of two banks. One compounds interest quarterly, the other daily. How much more will she earn with daily interest than with quarterly interest? The formula to compute the total amount T given principal P, interest rate R, time N and number of times interest is compounded J is:

$$T = P ° \left(1 + \frac{R}{J}\right)^{J * N}$$

THE DECISION BLOCK 2-6

Decision blocks provide a means for *conditional branching*, i.e., transferring to different blocks depending on certain conditions. For example, consider the flowchart in Figure 2.18.

The PAY depends on the number of hours worked. If HRS is greater than 40 an overtime pay is computed; on the other hand if HRS is less than or equal to 40 the regular PAY is computed. Hence HRS must be compared to 40.

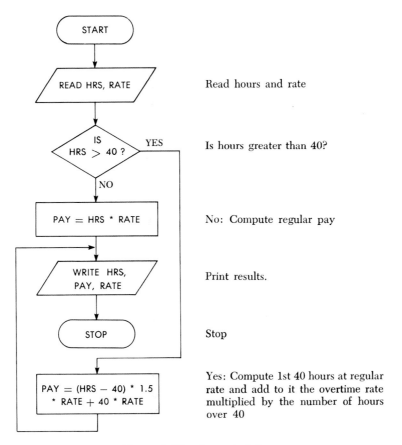

Read hours and rate

Is hours greater than 40?

No: Compute regular pay

Print results.

Stop

Yes: Compute 1st 40 hours at regular rate and add to it the overtime rate multiplied by the number of hours over 40

Figure 2.18.   Pay calculation

Conditional branches allow for by-passing of certain blocks in the flow-chart. In this way a flowchart can exhibit several alternate paths, which may be taken depending on whether or not certain conditions are met. All decisions must ultimately be made in terms of a comparison be-tween two variables or expressions. The following symbols are often used:

Symbol	Meaning
$<$	less than
$\leqslant$	less than or equal to
$=$	equal to
$>$	greater than
$\geqslant$	greater than or equal to
$\not<$	not less than
$\not>$	not greater than
$\neq$	not equal to

Example 1:

Two records are read. The first contains the amount due for a cus-tomer's charge account. The second contains the amount received from the customer. These two numbers are compared. If they are the same, the customer has paid his bill and owes nothing. If the amount received is greater than the amount due, the customer is given credit toward his account and has a credit balance. How-ever, if the amount received is less than the amount owed, the new balance is computed by adding a 2 percent service charge to the unpaid amount. The flowchart for this problem is shown in Figure 2.19.

Note the use of quotes around the message to be printed in the input-output block:

This indicates that the characters enclosed in quotes should be written on the output medium and should not be construed as variable names. Messages, headers and other types of literal information should always be enclosed in quotes in the WRITE statement. Note that the follow-ing two WRITE blocks produce different outputs.

WRITE
"PAY"

and

WRITE
PAY

In the first block the characters PAY (word) will be printed on the output form while in the second block the contents of memory location PAY will be printed out, for example 120.50.

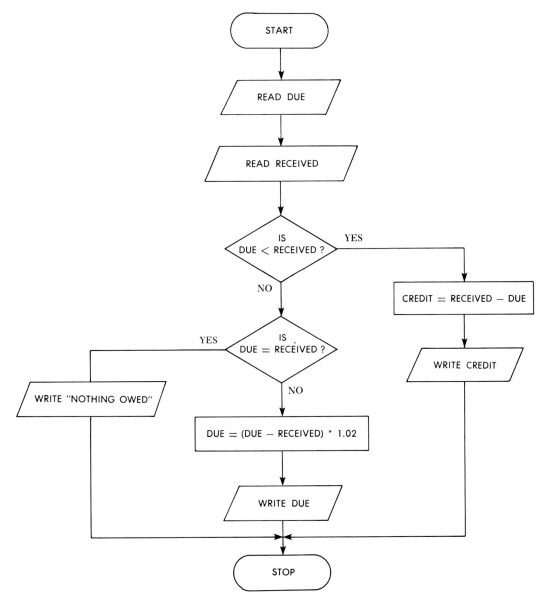

Figure 2.19.   Flowchart for customer billing problem

Example 2:

Suppose we are to read three numbers N1, N2 and N3 to determine the largest of the three numbers. Three decision statements are needed as shown in Figure 2.20.

Read the three numbers.

Is first > second?

NO

    Is second < third?
    YES
        LARG is third number

    NO
        LARG is second number
        go and print it

YES

    Is first < third?
    YES
        LARG is third number

    NO
        LARG is first number
        go and print it

Print LARG which happens to be the largest number

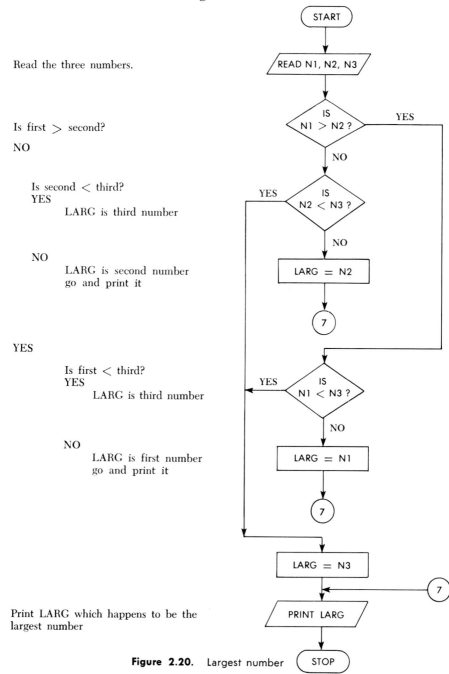

Figure 2.20. Largest number

EXERCISES 2-2

1. Trace the flowchart in Figure 2.19 for
   a. DUE = 50, RECEIVED = 100
   b. RECEIVED = 100, DUE = 150
   c. DUE = 50, RECEIVED = 50
2. A record consists of three fields: A name, a rate of pay and a number of hours worked. Draw a flowchart to produce the following output:

   > NAME
   > OVERTIME HOURS IS xxx
   > OVERTIME PAY IS xxx.xx
   > TOTAL PAY IS xxx.xx

   If the number of hours is less than 40 the two overtime entries in the output should be 0.
3. At Podunk University a student is enrolled in a course in which the final grade is either a "P" (Pass) or "F" (Fail). The grade is determined by adding the scores obtained on three tests: T1, T2, and T3. If the sum is greater than 225, a grade of "P" is assigned; otherwise, an "F" is assigned. Draw a flowchart to read a record containing three test scores and assign an appropriate grade.
4. Draw a flowchart to read a record containing two numbers and write out the larger of the two. Can you do this with only one output block?
5. Draw a flowchart to read a number X from a record and determine if $70 < X < 90$, i.e., determine whether X is in the open interval 70 and 90.
6. A salesman receives a commission of 10 percent on all sales if he has sold at least $10,000 worth of mechandise in a pay period, but only 8.5 percent if his sales are below $10,000. Draw a flowchart to read a record containing the amount of sales, compute the commission, and print the commission earned.
7. A salesman is assigned a commission on the following basis:

Sale	Commission
$ 00–$500	2%
over 500–$5000	5%
over $5000	8%

   Draw a flowchart to read a record containing an amount of sales and calculate the commission.

8. A student in DP101 has his/her account number in columns 1-4 and three test scores in columns 11-19. The student's average is based on his/her two best scores. Print the student's account number and his/her average on the line and the three test scores on another line as follows:

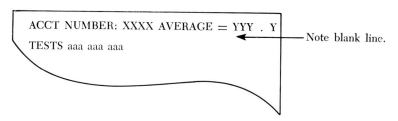

```
ACCT NUMBER: XXXX AVERAGE = YYY . Y
TESTS aaa aaa aaa
```
Note blank line.

9. Draw a flowchart to read a value for X and compute the absolute value of X. (Recall that $| X | = X$ if $X \geqslant 0$, $| X | = -X$ if $X < 0$).

10. Read three numbers from a record representing the three sides of a right triangle. Draw a flowchart to print which of these three numbers is the hypothenuse.

11. The FHA insures home mortgages requiring a down payment as follows:

   3% of the first $25,000
   5% of the remainder

The input consists of a record containing a social security number and a mortgage amount. Draw a flowchart to print the applicant's social security number and the amount of down payment required. Reject any application over $70,000.

12. Read an amount in cents between 0 and 100. Draw a flowchart to break down the amount read into the fewest number of quarters, dimes, nickels, and pennies as is possible: For example 76 cents = 3 quarters and 1 penny.

13. Jon Doe must decide whether to buy a house this year at relatively high interest rates or wait until next year when interest rates are anticipated to be lower but when inflation will have increased the cost of the house. This year he can buy a $50,000 house with 10% down and the balance financed at 15% for 30 years. Next year he believes he can buy the same house for $55,000 with 10% down and the balance financed at 14% for 30 years. Based on the total cost of the house (principal and interest) should he buy

now or wait? The formula to compute the amount of monthly payment M given the principal P, the interest rate R and the number of years T is given by:

$$M = \frac{P * \dfrac{R}{12}}{1 - \left(\dfrac{1}{1 + \dfrac{R}{12}}\right)^{T * 12}}$$

Draw a flowchart to make the decision for John Doe.

14. A certain metal is graded according to the results of three tests. These tests determine whether the metal satisfies the following specifications:

   a. Carbon content is below 0.67.
   b. Rockwell hardness is no less than 50.
   c. Tensile strength is greater than 70,000 psi.

   The metal is graded 1 if it passes all three tests, 9 if it passes only tests 1 and 2. It is graded 8 if it passes only test 1, and 7 if it passes none of the tests. Draw a flowchart to read a carbon content, a Rockwell constant, and a tensile strength, and determine the grade of the metal.

15. A telephone company charges its customers based on message units used each month according to the following schedule:

Units	Charge
up to 80	$7.00
81–100	$7.00 plus .03 for each unit over 80
101–120	$7.60 plus .025 for each unit over 100
121 and up	$8.10 plus .02 for each unit over 120

   Draw a flowchart to read the number of message units used and compute and print the telephone bill.

16. ACME Department Store charges its customers a finance charge on any unpaid balance based on a credit rating as follows:

Credit rating	Rate
1	1%
2	1.2%
3	1.4%
4	1.5%

Draw a flowchart to read a data record containing amount-owed, amount paid and credit rating. Compute and print the amount of service charge due. (Note that there is no service charge on an account that has a credit balance.)

17. Acme Rental, Inc. charges customers for use of its rental cars based on type of car, insurance purchased, mileage used, and number of days. The company leases three types of cars as follows:

Type	Daily charge	Mileage charge
1	$8	.06
2	$10	.08
3	$15	.12

The company offers two insurance plans:

Plan	Cost
1	20% of total daily charges; $10 minimum
2	$15 plus $3 per day of use

Draw a flowchart to read a data record containing type-of-car, insurance-plan, mileage, and number-of-days. Compute and print the various charges as well as the total.

18. Modify the flowchart for Exercise 17 above to include the provision that Insurance Plan 2 is not offered for Type 3 cars.

19. Modify the flowchart of Figure 2.20 to print the smallest number.

3 Loops

In all flowcharts discussed thus far, all branches pointed to blocks that had never been executed previously. Many times it is convenient to process blocks repeatedly. Suppose, for example, we wished to read and print the contents of two records. We could accomplish this operation by drawing the flowchart shown in Figure 3.1. Since the two sets of input/output blocks are identical it is possible to use only one set and execute it repeatedly as shown in Figure 3.2. This process of executing one or more blocks repeatedly is called *looping*. The group of blocks processed in this fashion is called the *body of the loop*.

END OF FILE CHECKS 3-1

When reading a file it is important that no attempt be made to read more records than are present in the file. Some means of avoiding an infinite loop in reading records must be implemented. This section discusses three end of file detection techniques: automatic end of file, trip record and last record code.

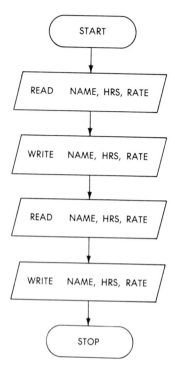

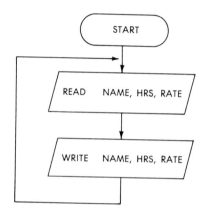

Figure 3.1. Reading and writing records without a loop

Figure 3.2. Reading and writing records with a loop

Automatic End of File

On some computer systems the computer can sense when there are no more records at the input device; thus the programmer can instruct the computer to keep reading data records until these run out. When the computer detects this condition (no more records) it will transfer to a predetermined location in the flowchart/program, specified by the programmer. This automatic end of file detection is very convenient and can be expressed in a flowchart by means of the special Input block shown in Figure 3.3.

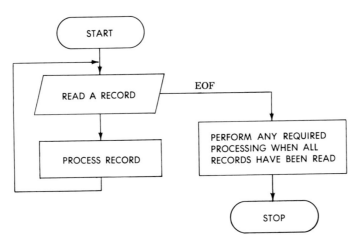

Figure 3.3. Automatic end of file check

In the case of Figure 3.3 the Input block has two exits. The EOF exit is taken when the computer senses that there are no more records to be processed at the input device. If there are still more records the other flowline is taken to process the record just read.

Example: Draw a flowchart to read a file with an unknown number ber of records each containing a grade, and provide a listing of all grades. A flowchart to solve this problem is shown in Figure 3.4.

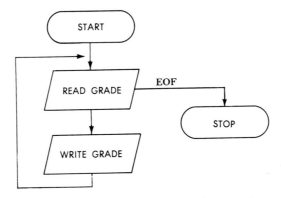

Figure 3.4. Automatic end of file example

Trip Record Method

Some computer systems do not support automatic end of file. It is then the programmer's responsibility to tell the computer system when it is reading the last data record. The flowchart must then be written to simulate an end of file condition. A special end of file code called a "trip code" purposefully different than any of the data items read is placed on the last record of the data file. Every time a record is read, the content of the data field is checked to determine if it contains the trip code.

Example: Draw a flowchart to read a data file with an unknown number of records; each record contains two grades. Print all grades, two per line. The flowchart to solve this problem is shown in Figure 3.5.

In the flowchart of Figure 3.5 G1 is used as a trip code. Every time a record is read, G1 is compared to zero. If G1 is not less than zero then the record read is not the last one and more records are read. If G1 is negative, reading is terminated and the last record (trip code) is not printed. Note that G2 could just as well have been chosen for the trip code.

The selection of a trip code depends on the nature of the data. Generally the data will fall within a certain range of values; a trip code can be any value outside the known range for the data values.

It should be emphasized that when using the trip record method for end of file condition, the test for trip code should immediately follow the READ statement, otherwise the trip code itself may be processed as part of the original data. Consider, for example, the flowchart in Figure 3.6.

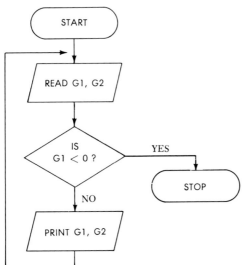

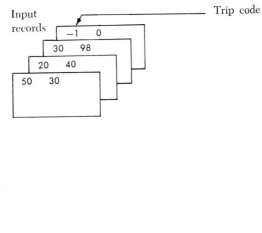

Figure 3.5.   End of file detection using a trip record

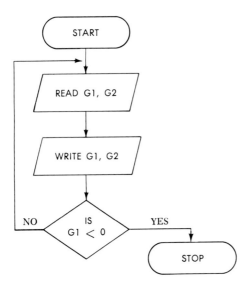

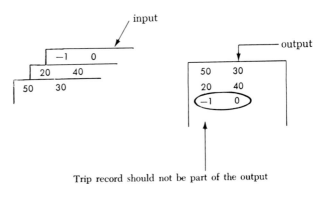

Figure 3.6.   Incorrect use of the trip record method

Last Record Code Method

The last record code method requires the programmer to set up his own end of file check. Consider for example that data file shown in Figure 3.7.

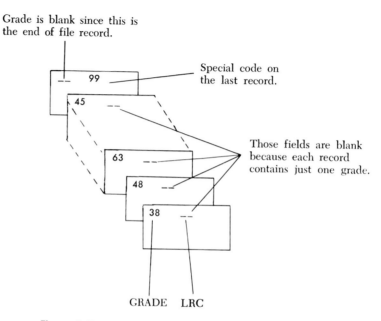

Figure 3.7.   File with user defined end of file record

Each record has two fields: GRADE and LRC (last record code). The field GRADE contains a student's grade. There is one record for each student. On each ordinary data record the field LRC is left blank; however, on the "end of file" record, the GRADE field is blank and the LRC field is set to 99. The special code in LRC designates that record as the last (trailer) record and is used to determine when the last record has been read. The flowchart shown in Figure 3.8 could be used to read the file shown in Figure 3.7 and to produce a list of grades.[1]

1. When using the last record code method, the check for last record code should immediately follow the READ statement, just as in the trip code method.

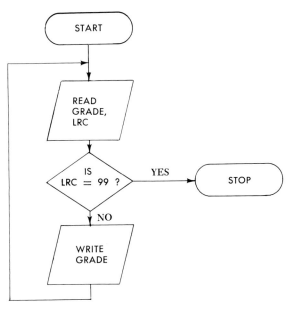

Figure 3.8.   A user defined end of file check

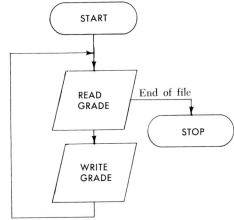

Figure 3.9.   A system end of file check

We will assume that if a field on a record is blank, it will be inter-preted as zero when read.[2] The flowchart shown in Figure 3.9 accomplishes the same purpose as Figure 3.8 except that a system end of file check is used.

It is recommended that in solving the problems in this book, the reader experiment with the various end of file checking methods discussed in this chapter.

2.   In some programming systems a blank field cannot be treated as though it contained the value zero. When using such a system all numeric fields must be entered with a numerical value zero.

3-2 COUNTING FOR LOOP CONTROL

The preceding section described means for avoiding the *infinite loop* illustrated in Figure 3.2. In that procedure there was no means to break away from the READ cycle. Exiting from a loop requires some decision within the body of the loop that results in a branch to some statement outside the loop. The nature of the decision to exit from a loop varies from one problem to another; sometimes exiting can be aided by a technique called *counting*. Some variable acting as a counter is set (initialized) to a starting value (for example, zero) before entering the loop. Each time the loop is executed, the value one is added to the counter (called *incrementing* the counter). The counter is then tested to determine whether the body of the loop has been executed a specific number of times; if this is not the case, the loop is executed again until the exit condition has been met at which time processing continues outside the loop.

Let us construct a flowchart to read a value from a record and write it out. We will process exactly 3 records. Counting in this case means keeping track of the number of times the input/output blocks are processed. This can be achieved by initializing a counter to zero, for example $\boxed{I = 0}$, before the first input block is processed. Following the output block, the counter is incremented by one in a block such as $\boxed{I = I + 1}$. In this way the variable I always reflects the number of times the input/output blocks are processed, thus indicating the number of records read and the number of values printed.

To read three records, a decision block such as $\langle \text{IS } I < 3 ? \rangle$ is required after incrementing the counter. So long as the counter is less than three, the loop is executed again. When I becomes three, three records will have been processed and the exit is taken. A complete flowchart to process three records is shown in Figure 3.10.

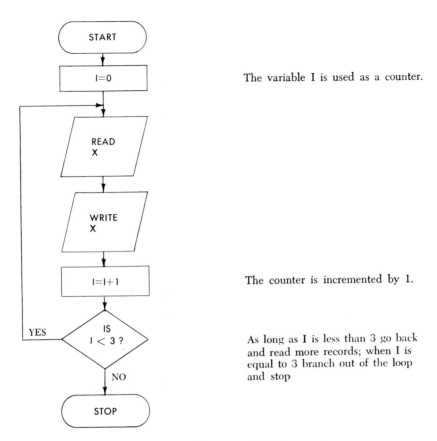

The variable I is used as a counter.

The counter is incremented by 1.

As long as I is less than 3 go back and read more records; when I is equal to 3 branch out of the loop and stop

Figure 3.10. A loop controlled by counting

A useful technique for following the logic of a flowchart is to create a table displaying the values of each variable after execution of each block in the flowchart for a sample of input records. For example, the execution of the algorithm of Figure 3.10 may be described by means of a table such as that shown in Figure 3.11. Recall that the value of a variable is destroyed when a new value is assigned to that variable. The variable X, in this case, assumes three different values; however, each new value replaces the preceding one. A shorter and more practical arrangement of the table is shown in Figure 3.12.

I	X	Value written	Comment
0			Results from I=0.
	38		From READ X (the first data record).
		38	X is printed.
1			The value of the expression 0+1 is placed in I.
			I is compared to 3. I is less than 3, therefore another record is read.
	45		From READ X (the second data record).
		45	X is printed.
2			I+1 is placed in I; therefore the new value for I is 2.
			I is compared to 3; 2 is less than 3, therefore another record is read.
	36		From READ X (the third data record).
		36	X is printed.
3			I=I+1; 2+1 is 3.
			I is compared to 3; 3 is not less than 3, therefore STOP.

Sample group of input records.

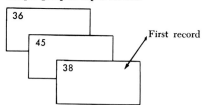

First record

Figure 3.11. Execution of the algorithm in Figure 3.10 with a sample group of input records

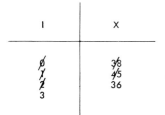

I	X
~~0~~ ~~1~~ ~~2~~ 3	~~3~~ ~~8~~ ~~4~~ ~~5~~ 36

Figure 3.12. A table for tabulating values assumed by variables

Many times a file may consist of a variable number of records. Counting can be used to control the processing of such files if an initial record is read containing the number of records which follow. The counter is then compared to that number to determine when to exit from the loop. See, for example, the flowchart in Figure 3.13. With the sample data shown, three records will be read and the value contained on each will be listed on the output device.

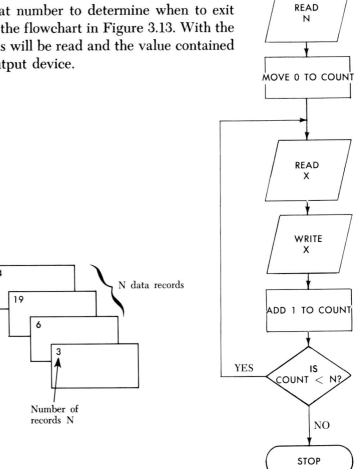

Figure 3.13. Processing a variable number of records

EXERCISES 3-1

1. Use the tabular method to trace the values assumed by the variables for the flowchart in Figure 3.13 with the sample data shown.
2. Compare the flowcharts in Figure 3.14. How many records will be read by each? Is it necessary that the initial value of a counter be zero? If the initial value is one, how is the test changed?
3. Does the flowchart in Figure 3.15 represent an algorithm for reading three records. Why or why not?
4. How many records will be read by each of the flowcharts in Figures 3.16a, 3.16b, and 3.16c?
5. Consider the flowchart in Figure 3.17. How many times will the READ instruction be executed? Is it necessary to begin a counter with the value one or zero? How is the test changed?

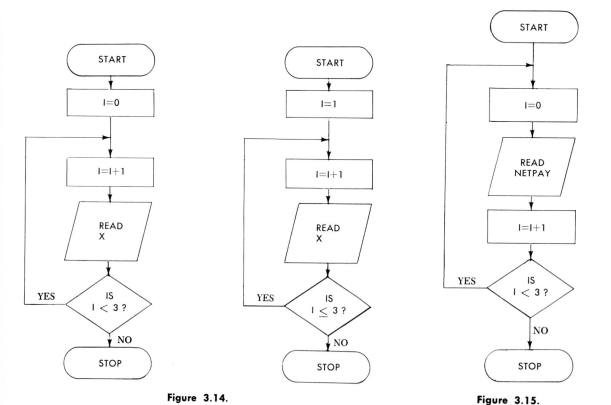

Figure 3.14.   Figure 3.15.

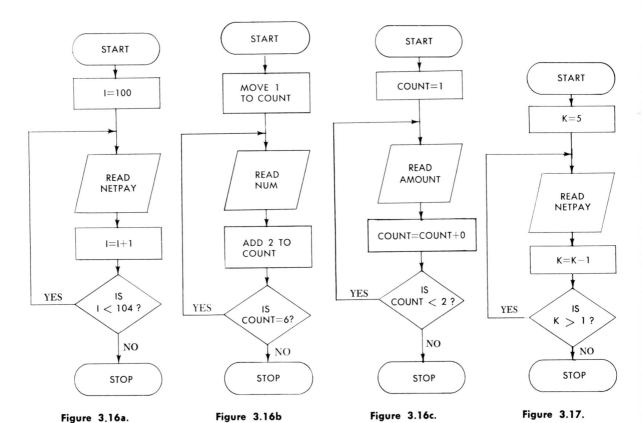

Figure 3.16a.   Figure 3.16b   Figure 3.16c.   Figure 3.17.

6. Construct a table similar to that in Figure 3.12 for the flowchart and records in Figure 3.18. How many values will be read for SALES?

7. The XYZ Company must pay ten men for a special project. Each of the ten employee records contain the hours worked, the rate of pay, and the total amount to be withheld. Draw a flowchart to prepare a payroll report showing all input data and the amount of each paycheck.

8. Repeat Exercise 7 but process a variable number of records by reading an initial record containing the number of data records to follow.

9. A data file contains an unknown number of pairs of grades, one pair per record. Draw a flowchart to compute and print each pair of grades and the sum of each pair.

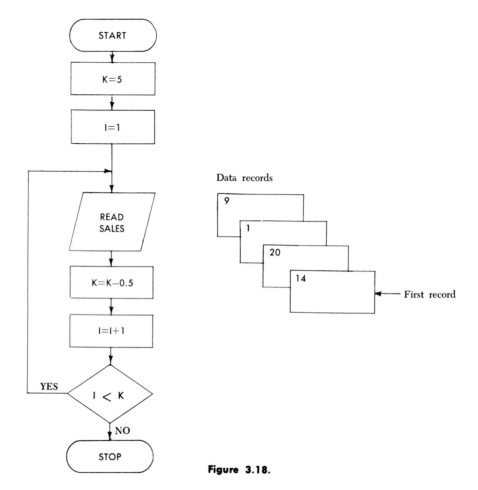

Figure 3.18.

10. Draw a flowchart to process a data file containing loan applicant data as described in Exercises 2-2 Number 11. Use any end of file technique you desire.
11. Draw a flowchart to process the data file described in Exercises 2-2 Number 15. Assume that the last data record has −1 in the message units field.
12. Each record of an input file consisting of 10 records contains three entries: a name, a rate of pay and a number of hours worked. Write a program using the counting method, to compute each employee's pay. Overtime is paid at time and a half regular rate and *all* employees earning over $500 get a bonus equal to 1.8 percent of their gross pay. The output should be similar to the following:

Input data

DOE	5	40
LIKE	20	60
SAM	10	50
MAT	5	50

Output

NAME	RATE	HOURS	PAY
DOE	5	40	200
LIKE	20	60	1425.2
SAM	10	50	559.9
MAT	5	50	275

13. Draw a flowchart to process the data file described in Exercises 2-2 Number 16. Assume that the credit rating is 0 on the last record.

14. Draw a flowchart to process the data file described in Exercises 2-2 Number 17. Make an appropriate assumption about the last record.

15. A data file contains an initial record specifying the number of data records following. Each data record contains the following fields: employee number, years experience, position code, weekly-pay. Each employee is to be assigned a Christmas bonus based on the following rules:

Position-code	Bonus
1	one week's pay
2	two weeks pay, maximum of $700
3	1½ weeks pay

Employees with more than 10 years experience are to receive an additional $100; Employees with less than 2 years experience receive half the usual bonus.

Draw a flowchart to process this data file and calculate Christmas bonuses.

16. Modify the flowchart for Exercises 2-2 Number 7 to process a complete data file. What assumption can be made about the last record?

MORE ON COUNTING 3-3

The technique of counting is useful in a variety of ways other than loop control. Consider for example, the following problems:

Example 1:

A file contains an unknown number of records. To determine the number of records in the file (exclusive of the end of file record) we initialize a counter to zero prior to entering the input loop, and increment the counter by one each time the READ instruction is executed. When the end of file is encountered, the value of the counter will reflect the total number of data records read. (See Figure 3.19)

Example 2:

Each record of a data file contains a student's grade. It is desired to count the number of grades that are greater than N (where N is a variable read initially), and calculate the percentage of grades greater than N. A grade outside the range 0-100 will terminate input. A flowchart for this problem is shown in Figure 3.20. Note how the trip record test is performed immediately after reading each grade.

Example 3:

Each record in a file contains three fields: age, sex, and marital status as described in Figure 3.21. It is desired to count the number of single males under 25 years of age using the last record

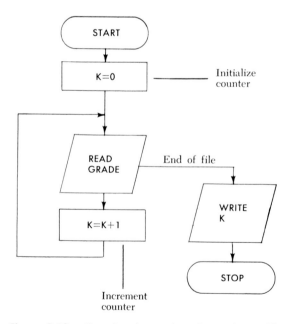

Figure 3.19. Counting the number of records in a file

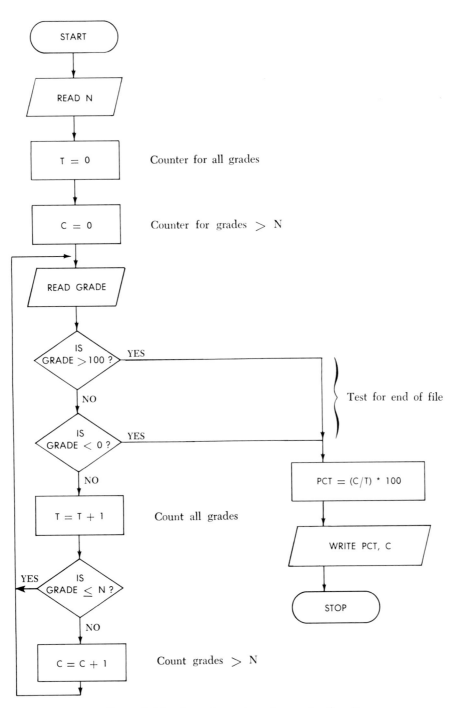

Figure 3.20. Percentage of grades greater than N

code method to stop reading the file. The flowchart for this problem is shown in Figure 3.22. Note again that the check for last record code should be done after reading each record, otherwise AGE, SEX and STATUS will all be zero for the last record.

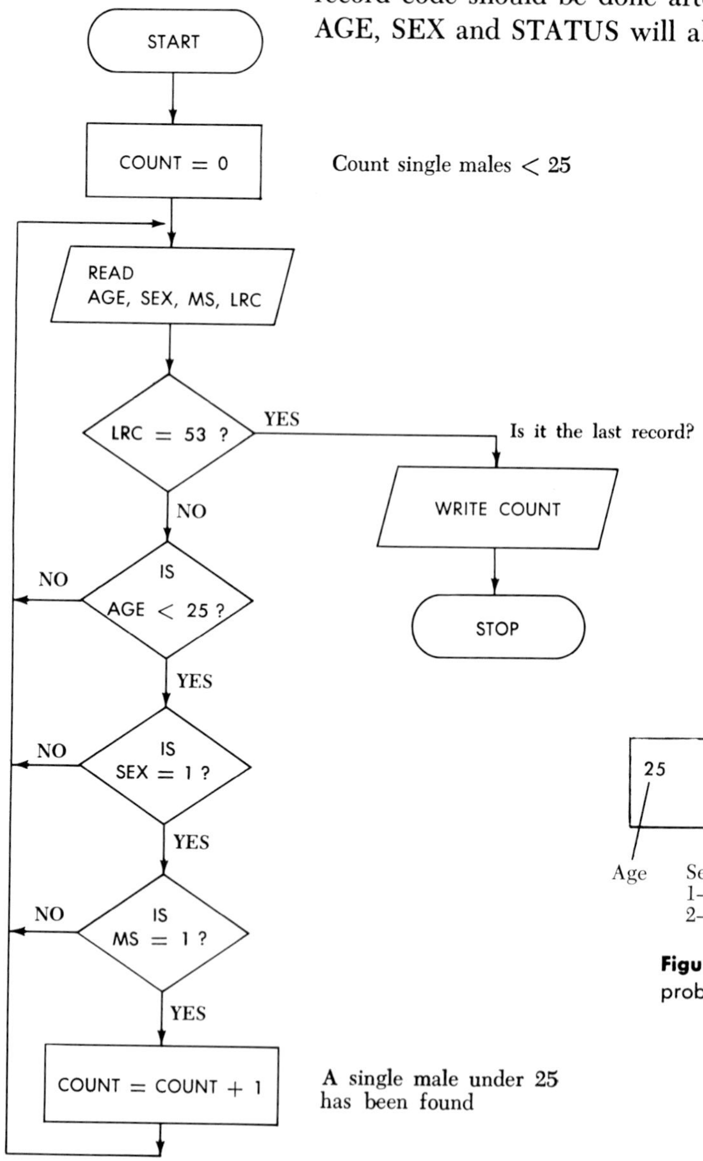
START
COUNT = 0
Count single males < 25
READ
AGE, SEX, MS, LRC
LRC = 53 ?
YES
Is it the last record?
NO
WRITE COUNT
STOP
IS
AGE < 25 ?
NO
YES
IS
SEX = 1 ?
NO
YES
IS
MS = 1 ?
NO
YES
COUNT = COUNT + 1
A single male under 25 has been found

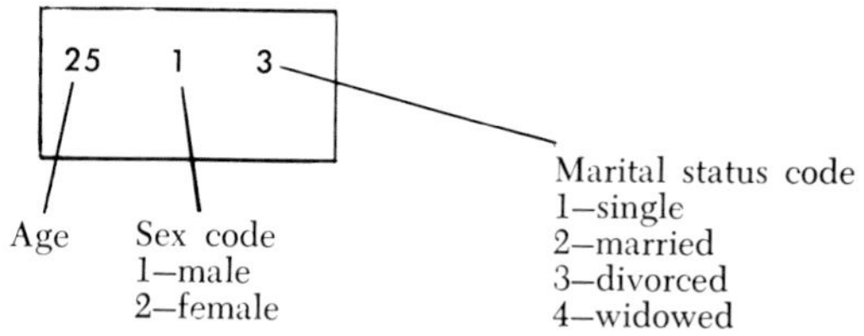
25 1 3
Age
Sex code
1—male
2—female
Marital status code
1—single
2—married
3—divorced
4—widowed

Figure 3.21. Data record for counting problem

Figure 3.22. Flowchart for counting problem

Example 4:

Draw a flowchart for a program to produce a table showing the Celsius equivalent for Fahrenheit temperatures from 0° to 212°. The required flowchart is shown in Figure 3.23.

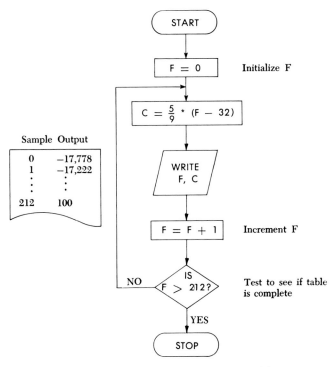

Figure 3.23.   Fahrenheit to Celsius Table

Example 5:

Another important application of counting is number generation. For example, to generate the even positive integers, the statement $I = I + 2$ is repeatedly executed with I set initially to 0. To generate the odd-positive integers, the same formula could be used, with I initially set to 1. The flowchart shown in Figure 3.24 prints a 12's multiplication table. The numbers 1, 2, 3, . . . , 10 can be generated by the statement $I = I + 1$, with I initially set to 0. There is no need to read these numbers from input records, since they can be easily generated. Each time a new value is generated for I, it

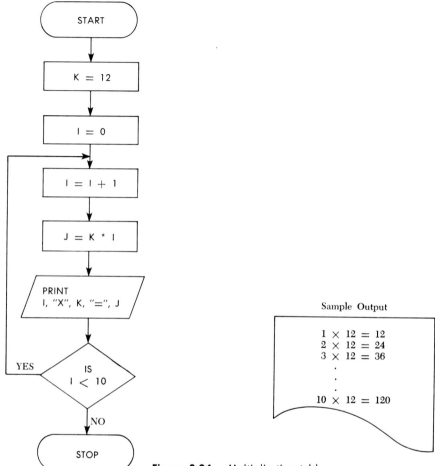

Figure 3.24.   Multiplication table

is immediately included in the computational formulas K*I. The value of I is then compared to 10 to determine the end of the program. A program to solve this problem is shown in Figure 3.24.

Example 6: Search for a Largest Grade

Each record of a set of records contains a grade (0-100). It is not known how many records there are in the set. We want to draw a flowchart to print the highest grade. Since we do not know how many records there are, we could add to the set of records a last record with a negative grade on it, so that every time we read a record we can ask the question, "Is the grade read negative?" If it is, it means the end of the set has been reached; otherwise more records need to be read, and processed.

To determine the largest grade requires that we compare successively a new grade with the highest grade found so far (MAX); if the new grade read is larger than MAX, we replace MAX by the new grade; otherwise we keep on reading grades until we find one that is larger than MAX (if there is one). At the end of the program, MAX is the highest grade. To start the comparisons we can set MAX equal to the first grade (after all, if there were only one grade, MAX would be the highest score!). A flowchart with accompanying tabulation is shown in Figure 3.25.

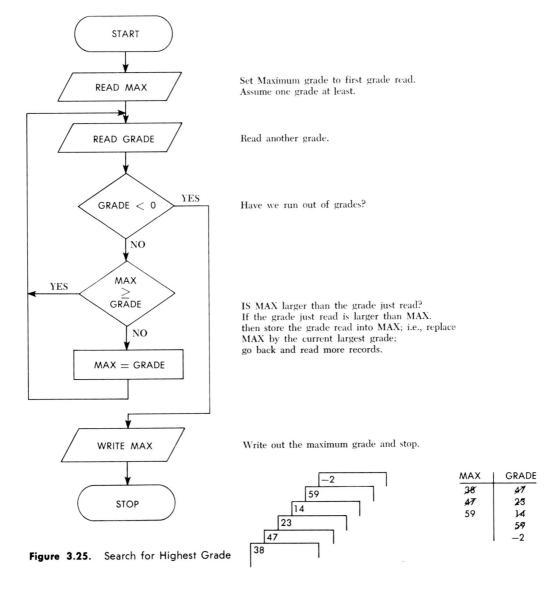

Figure 3.25.   Search for Highest Grade

EXERCISES 3-2

1. Consider the file shown in Figure 3.26 and the flowcharts in Figures 3.27a and 3.27b.

    Determine the number of records read and the final value for K.

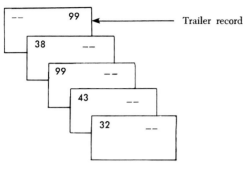

Trailer record

Figure 3.26.

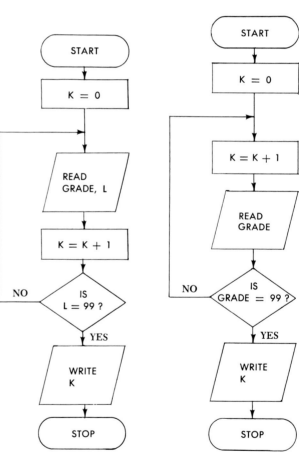

Figure 3.27a.   **Figure 3.27b.**

2. Tabulate the contents of each variable in the flowchart shown in Figure 3.22 for the following file:

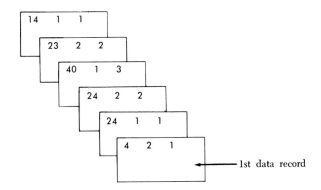

3. Modify the flowchart of Figure 3.23 to produce the table with Fahrenheit temperatures varying from 212° to 0°.

4. Draw flowcharts to produce each of the following sequences:
   a. 2, 4, 6, 8, . . . 100
   b. 1, 3, 5, 7, 9, . . . 99
   c. 5, 10, 15, 20, . . . 100
   d. 1, 10, 100, 1000, . . . 1,000,000,000
   e. 2, 4, 8, 16, 32, . . . 4096
   f. 100, 99, 98, . . . 1
   g. 3, 8, 13, 18, . . . 48
   h. −1, 1, −1, 1, . . . −1 (25 terms)
   i. $\dfrac{1}{2}, \dfrac{1}{3}, \dfrac{1}{4}, \dfrac{1}{5}, \cdots \dfrac{1}{30}$
   j. $\dfrac{1}{2}, \dfrac{-1}{4}, \dfrac{1}{6}, \dfrac{-1}{8}, \cdots \dfrac{1}{50}$
   k. $\dfrac{1}{9}, \dfrac{1}{99}, \dfrac{1}{999}, \cdots \dfrac{1}{999,999}$

5. Draw a flowchart to produce a table showing the circumference C and the area A of circles with radii R varying from 1 to 30. (Note $A = \pi R^2$, $C = 2\pi R$)

6. Draw a flowchart to produce a table showing the amount of monthly payment required for a loan with interest rates varying from 10% to 18%. Use the formula given in Exercises 2-2 Number 13.

7. Modify the flowchart of Exercises 2-2 Number 12 to produce a table for amounts varying from 1 to 100 cents.

8. Each record in a file contains a student's grade. Draw a flowchart to determine the number of grades less than 53.

9. Each record in a given file contains one number. Calculate and print the number of negative, zero, and positive values in the file.

10. Draw separate flowcharts to read a file having the same format as in Figure 3.21 to calculate each of the following:
    a. The percentage of students over thirty years of age
    b. The percentage of students who are not married
    c. The number of students who are either widowed or divorced
    d. The number of students who are under thirty years of age but more than twenty

11. Each day a record is prepared with the total amount of sales for that day. Draw a flowchart to determine the number of days in which the total sales was below $1,000.

12. Each record of a data file contains a social security number, a number of hours and an hourly rate of pay. Draw a flowchart using the trip record method to print the social security numbers of those employees having worked more than 40 hours, the number of overtime hours they worked, and their total pay. Also print the number of employees having worked overtime. Overtime is calculated at one and one-half times regular pay.

13. Each record of an input file contains an age and a sex code (Female = 1, Male = 2). Draw a flowchart to compute and print the percentage of females under 30 (out of the total female population). The output should be similar to:

AGE	SEX
60	MALE
45	FEMALE
23	MALE
41	FEMALE
18	FEMALE
29	FEMALE

PERCENTAGE OF SPECIAL
FEMALES IS 50 PERCENT

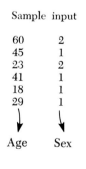

Sample input

60	2
45	1
23	2
41	1
18	1
29	1

Age   Sex

14. Records have been prepared for a store showing daily sales for corresponding days of succeeding years. The first value on the record is the amount of sales for a given day in the first year; the second value is the amount of sales for the corresponding day in the second year. Find the number of days in which the second year's daily sales exceeds the first year's daily sales by more than 10 percent of the first year's sale. The output should be similar to the following:

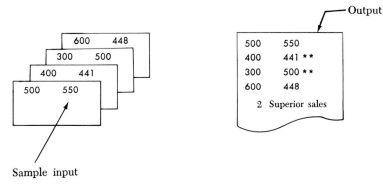

Sample input

Note that "**" identify those records for which the second year's daily sales exceeds the first year's daily sales by 10%.

15. There are 8,000 alumni at Twidlee Dum University. For each alumnus there is a record containing a social security number, year of graduation, and college within the University. Colleges are coded as follows:

1	Sciences
2	Liberal Arts
3	Engineering
4	Business

The class of 1964, college of sciences is to have a reunion. The chairman of the event needs a list of social security numbers of alumni of this class and college and the total number of alumni in the group. Supply this information.

16. A file consists of an unknown number of records; each record contains a social security number, a sex code (1 = male, 2 = female), and an earning. Draw a flowchart using the sex code

as a trip code to count the number of males earning over $30,000 and a list of females by social security number earning less than $10,000.

17. Each record in a file contains one number. These numbers are supposedly arranged in ascending numerical order. Draw a flowchart to determine whether the numbers are indeed sorted into ascending sequence. a) Print an error message and the record number of the first out of sequence value if the input is not in ascending sequence. b) Print all input numbers with an appropriate message; for example:

> 15
> 25
> 8 out of sequence
> 30
> 26 out of sequence

18. A file consists of an unknown number of records, each containing two numbers. Count how many positive, zero and negative numbers there are, using the automatic end of file for input termination. The output should be similar to the following, given the sample input file.

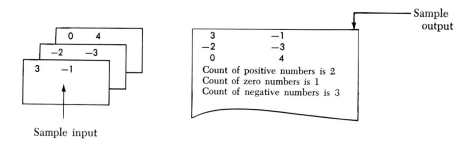

Sample input

Sample output

19. Write a program to print out multiplication tables from 2 to 12 as follows:

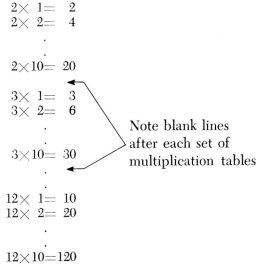

$$2\times 1= \quad 2$$
$$2\times 2= \quad 4$$
$$.$$
$$.$$
$$2\times 10= \quad 20$$

$$3\times 1= \quad 3$$
$$3\times 2= \quad 6$$
$$.$$
$$.$$
$$3\times 10= \quad 30$$
$$.$$

$$12\times 1= \quad 10$$
$$12\times 2= \quad 20$$
$$.$$
$$12\times 10=120$$

Note blank lines after each set of multiplication tables

20. Mrs. X has just invested $9,000 at 13.5 percent yearly interest rate. She has decided to withdraw the accumulated interest as soon as that interest has exceeded $11,000.00. How many years will Mrs. X have to wait before she can withdraw at least $11,000 of accumulated interest?

Could you be more specific and identify the number of years and months?

THE ACCUMULATION PROCESS 3-4

We have already seen how counting is made possible by repeated execution of such statements as $I = I + 1$ where I is initially set to a starting value (for example $I = 0$). Each time the statement $I = I + 1$ is executed, the value 1 is added to the counter I which takes on successive values 1, 2, 3, 4, and so on.

Counting can be thought of as "accumulating" a count. The main difference between "counting" and "accumulating" is that instead of adding repetitively a constant (1 for example) to a counter, a variable is added or multiplied repetitively to an accumulator. The accumulator is itself a variable used to keep track of partial or running sums, products, etc (see Figure 3.28).

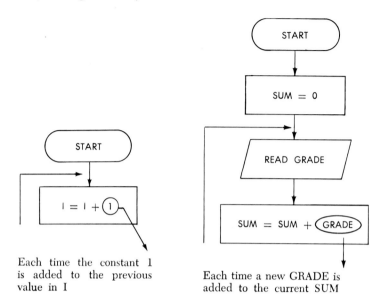

Each time the constant 1 is added to the previous value in I

Each time a new GRADE is added to the current SUM

Figure 3.28. Counting and Accumulating

The accumulation process may be described more generally as the repeated execution of statements of the form

$$\text{Variable} = \text{Expression}$$

where the variable on the left side (called an accumulator) is also present in the expression on the right side. The following are examples of such statements:

a) SAM = 2 ° SAM
b) SUM = SUM + GRADE
c) PROD = PROD ° N

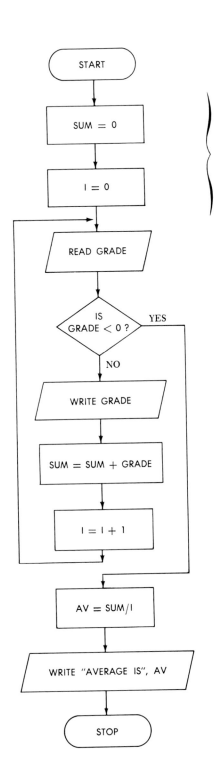

Initialization process

SUM is used as an accumulator to add all grades

I is used to count records (grades)

Read a grade

Is it the end of the file?

NO: Print it and add the grade just read to the SUM and count the number of records (grades)

YES: All grades have been read; SUM contains the sum of all grades exclusive of the trip value. I is the number of grades. AV is then the average of all grades.

Figure 3.29. Average of an unknown number of grades

In example a) the statement SAM = 2 ° SAM really says "replace SAM by twice its value." If the value of SAM is initially 2, repeated execution of the statement would cause SAM to take on the values 4, 8, 16, 32, . . successively. To better understand the accumulation process consider the following two problems:

Problem 1: A data file contains an unknown number of records. Each record contains a grade; a negative grade identifies the end of the file. Draw a flowchart to print each grade and the grade average. The solution to this problem is shown in Figure 3.29:

The accumulation process in the flowchart of Figure 3.29 can be better understood or visualized if the successive values of SUM, I and GRADE are tabulated given a sample input as shown in Figure 3.30.

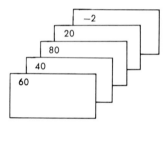

SUM	I	GRADE
0	~~0~~	~~60~~
~~60~~ (0 + 60)	~~1~~	~~40~~
~~100~~ (60 + 40)	~~2~~	~~80~~
~~180~~ (100 + 80)	~~3~~	~~20~~
200 (180 + 20)	4	−2

Figure 3.30. Tabulation of the flowchart of Figure 3.29

Problem 2: Draw a flowchart to compute the product of the first N positive integers (N! read as N *factorial*) where N is a positive integer read from one record. In this problem only the value for N is to be read. The numbers 1, 2, 3, 4, . . . N can be generated internally by the formula I = I + 1 (I initially set to 0); each time a new value is generated for I, it is multiplied to the product accumulator PROD as shown in Figure 3.31.

The accumulation process of the factorial function in Figure 3.31 can perhaps best be visualized through a specific example shown in Figure 3.32 for a value of N equal to 4. Note that only one record is read.

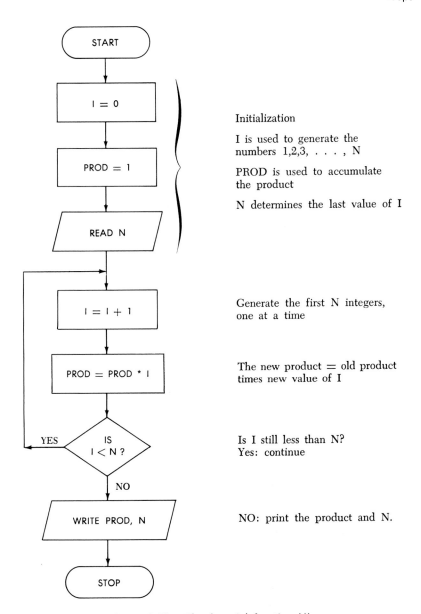

Initialization

I is used to generate the numbers 1,2,3, . . . , N

PROD is used to accumulate the product

N determines the last value of I

Generate the first N integers, one at a time

The new product = old product times new value of I

Is I still less than N? Yes: continue

NO: print the product and N.

Figure 3.31.   The factorial function N!

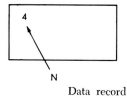

Data record

PROD	I	N
~~1~~ (1 × 1)	~~0~~	4
~~2~~ (1 × 2)	~~1~~	
~~6~~ (2 × 3)	~~2~~	
24 (6 × 4)	~~3~~	
	4	

Figure 3.32. Tabulation of the factorial function

3-5 CONNECTOR BLOCKS

Our flowcharts thus far have been sufficiently simple that all connecting flowlines could be drawn without crossing each other or without being too difficult to follow. To avoid confusion in more complex flowcharts, the connector block is used. For example, consider the two flowcharts shown in Figure 3.33. Flowchart A is drawn with solid flowlines; flowchart B is equivalent to flowchart A but drawn with connectors. The symbol within the connector serves two purposes: (1) to identify a block by a label for future reference, and (2) to indicate the target for a transfer. A *transfer* is indicated when the flowline points toward the connector block as in →(Z) and →(4.) The point to which transfer is made (an *entry* point) is indicated by the flowline pointing away from the connector as in (Z)→ and (4)→. If a transfer is made to another page of multipage flowchart, the symbol ⬡ (called an *off page connector*) may be used to mark transfer and entry points. Any number of transfers may be made to a single entry point; however, each entry point must be uniquely defined. For example, any number of transfers such as →(B) (B) may be made, but only one entry point (B)→ may be specified.

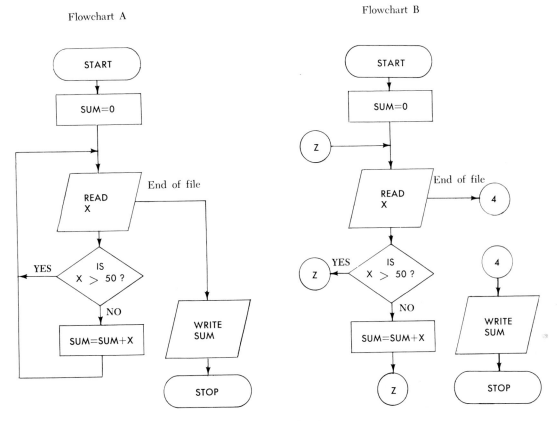

Figure 3.33. Example of connectors

EXERCISES 3-3

1. In the case of the factorial computation in Figure 3.31 why is it necessary to set PROD equal to 1 initially?
2. Consider the flowcharts shown in Figure 3.34a and 3.34b. Which would compute the average of five grades read from five data records?
3. Draw a flowchart to accumulate and print each of the following:
   a) $1 + 3 + 5 + 7 \ldots + 225$. Is accumulation necessary in this problem?
   b) $1^2 + 2^2 + 3^2 + 4^2 \ldots + 100^2$
   c) $2 * 4 * 6 * 8 \ldots * 100$
   d) $2 + 4 + 8 + 16 + 32 \ldots + 1024$
   e) $1 + 1/2 + 1/3 + 1/4 \ldots + 1/100$
   f) $1 - 2 + 3 - 4 + 5 \ldots - 100$

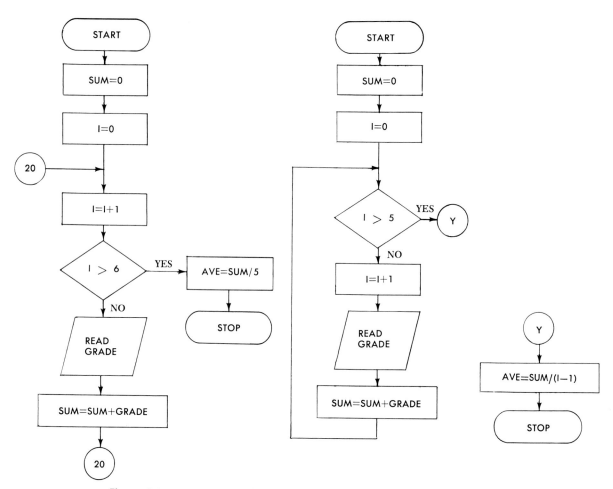

Figure 3.34a.

Figure 3.34b.

4. Draw a flowchart to read a positive value for J and print:

   a) the even positive integers less than or equal to J
   b) the sum of the first J odd integers

5. Consider the flowchart and sample file shown in Figure 3.35. What will be the value of LARG after execution of the algorithm?

6. Suppose it is known that the range of possible values for NUM in Figure 3.35 is zero to one hundred. Is it possible to determine the largest number with only one input block? (What value would have to be placed into LARG initially?)

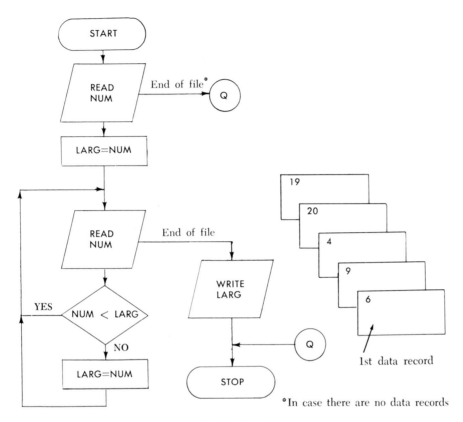

Figure 3.35.

7. Consider the flowchart and sample file shown in Figure 3.36. Tabulate the values assumed by the variables in execution of the algorithm. State in words the problem being solved.

8. John Doe has invested $1,000 at 8 percent interest for 10 years to be compounded annually. Draw a flowchart to calculate and write out the value of his investment at the end of each year.

9. How is the flowchart of Exercise 8 changed if interest is compounded quarterly?

10. Bon Bon's, Inc., a large department store, is having a sale on women's $29.98 hand bags. On each day of the sale the previous day's selling price will be reduced by 10 percent. The sale will last five days. Draw a flowchart to produce a table showing each day's sale price.

11. Write a program to read a value for N and compute the sum of the squares of the first N even integers. For example, if $N = 4$, the sum is $2^2 + 4^2 + 6^2 + 8^2$.

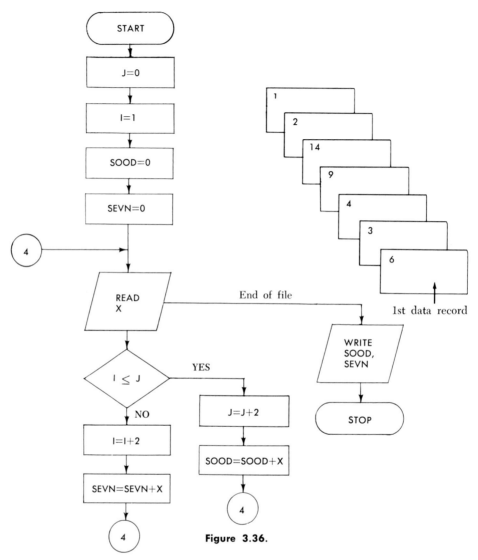

Figure 3.36.

12. Mr. X is hired for 30 days' work. The first day he earns $ .01; the second day, $ .02; the third day, $. 04; in general, double the previous day's earnings. Print out a table showing the day and the amount earned for each of the 30 days. Do not ask for the name of the company for which Mr. X works. The last line of the output should read

    TOTAL WAGES ARE $10737418.23 FOR 30 DAYS WORK

13. It can be shown that the irrational number $e = 2.71828 \ldots$ can be approximated by taking as many terms as desired in the relation

$$e = 1 + 1/1! + 1/2! + 1/3! + 1/4! + \ldots$$

Draw a flowchart to approximate the value of e using this method. Stop when successive approximations differ by less than .001.

For example: 1st approximation is 1

2nd approximation is $1 + 1/1! = 2$

3rd approximation is $1 + 1/1! + 1/2! = 2.5$

4th approximation is $1 + 1/1! + 1/2! + 1/3! =$ 2.666, etc.

14. Consider the flowchart and sample file shown in Figure 3.37.
    a. What statement is missing from the flowchart?
    b. After reading the record what value will be written for A?
    c. After reading the second record, what value will be written for A?

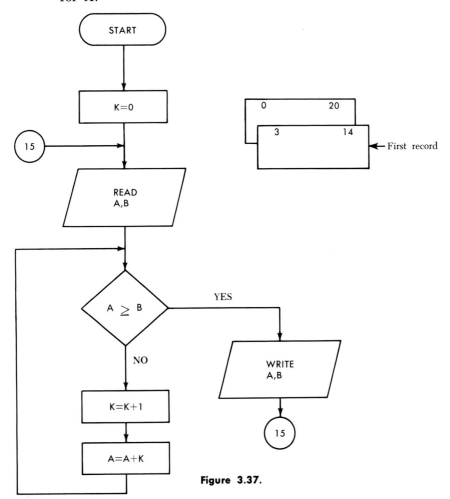

Figure 3.37.

15. A file contains an unknown number of records. Each record contains two numbers. Draw a flowchart to compute the sum of the positive numbers and the count of the negative numbers. The number 0 is considered positive. Each READ statement should specify two variable names. The output should be similar to the following given the sample input:

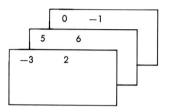

−3	2
5	6
0	−1

POSITIVE SUM = 13
COUNT NEGATIVE = 2

16. Every time a meal is sold at Charlie's Eatery a record is prepared with the cost of the order and a meal code: 1—breakfast, 2—lunch, 3—dinner. Draw a flowchart to print the input data and to compute:

    a) Total day's sales
    b) average cost of breakfasts
    c) minimum dinner cost

17. A file contains an unknown number of records (assume N records). Each record contains a grade. Draw a flowchart to compute the standard deviation of the grades. The formula for the standard deviation is given by:

$$S = \sqrt{\frac{N^* (X_1^2 + X_2^2 + X_3^2 \ldots X_N^2) - (X_1 + X_2 + X_3 \ldots X_N)^2}{N^*(N-1)}}$$

where $X_1, X_2, X_3, \ldots X_N$ are the 100 grades.

3-6 AUTOMATIC LOOP CONTROL

The automatic loop control represents no new flowcharting concepts, in fact any flowchart can be written the automatic loop control. The purpose of this feature is strictly one of coding convenience for the programmer. The automatic loop control block allows the user to control a loop using only one symbolic block instead of the three blocks generally required to execute a procedure a specific number of times. The automatic loop control replaces the initialization, the incrementation and the testing of the counter. The automatic loop control block is implemented in the various programming languages and may vary here and there. Figure 3.38 illustrates the use of the automatic loop control block to compute the average of ten grades read from ten records.[3]

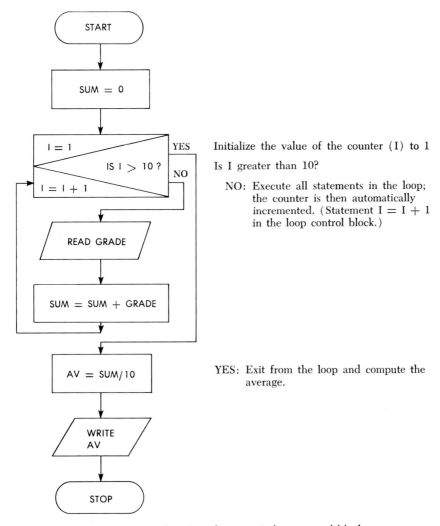

Initialize the value of the counter (I) to 1

Is I greater than 10?

    NO: Execute all statements in the loop; the counter is then automatically incremented. (Statement I = I + 1 in the loop control block.)

YES: Exit from the loop and compute the average.

Figure 3.38. Use of automatic loop control block

3. In some programming languages testing the counter may be performed at the end of the loop rather than at the beginning.

The automatic loop control block in Figure 3.38 is equivalent to the flowchart shown in Figure 3.39.

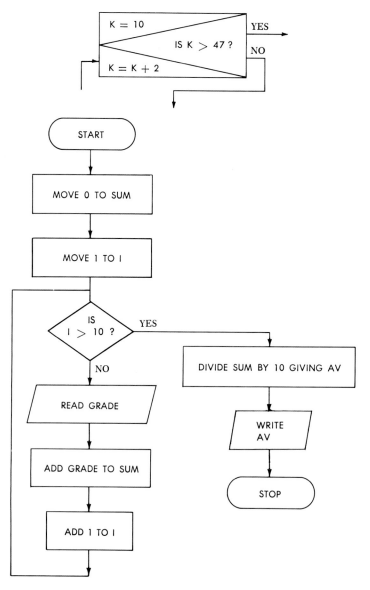

Figure 3.39. Flowchart equivalent of automatic loop control

Note that it is possible to initialize and increment the counter in the automatic loop control block in many different ways. In the following case, the statements in the body of the loop will be executed for values of K = 10, 12, 14, 16, . . . 44, 46.

EXERCISES 3-4

1. Each record in a file (6 records in all) contains a number representing the radius of a circle. Draw a flowchart to compute and print the circumference and area of each circle.
2. Draw a flowchart to create the 3's multiplication table up to 12.
3. Same as exercise 2 with tables ranging from 1 to 12.
4. Could you still use the automatic loop control to read a file with an unknown number of records assuming there are at most 500 records?
5. Rewrite the flowcharts for Exercises 3-2 Number 4 using automatic loop control.
6. Rewrite the flowcharts for Exercises 3-2 Numbers 5-7 using automatic loop control.

4 File Processing

Many computer applications require processing and storing large amounts of data. Data is organized into records; one record may contain information about one person, product, account, or other entity. A group of related records constitutes a *file*. For example, a retail store maintains an accounts receivable file. Each record in the file represents purchases and payments made by a particular customer. A manufacturer maintains an inventory file. There is one record in the file for each type of product manufactured. Other common files maintained by commercial concerns include payroll, personnel, accounts payable and so forth.

Information within a record is divided into *fields*, each with a specific type of data item recorded in it. For example, a record within a payroll file might contain such information as the employee's social security number, name, position, rate of pay, number of dependents, and marital status. Within a file all records usually have the same layout. The size of each field is constant regardless of the data placed in the field. For example, the name field in the layout in Figure 4.1 is 20 characters in length. A name is placed in the left portion of the field (*left justified*). Any unused positions remain blank. The rate of pay field is 4 characters in length, thus the largest pay rate that can be accommodated is $99.99. Should the rate of pay be a value taking fewer than four digits it would be entered into the field with leading zeros to accommodate field size (*right justified*). For example, the

Social Security Number	Name	Position	Rate of Pay	Dependents	Status	
9 9 9 9 9 9 9 9 9	9 9 9 9 9 9 9 9 9 9 9 9 9 9 9 9 9 9 9 9	9 9 9 9 9 9 9 9 9 9	9 9 9 9	9 9	9 9	9 9 9 9 9 9
1 2 3 4 5 6 7 8 9	10 11 12 13 14 15 16 17 18 19 20 21 22 23 24 25 26 27 28 29	30 31 32 33 34 35 36 37 38 39	40 41 42 43	44 45	46	47 48 49 50 51 52

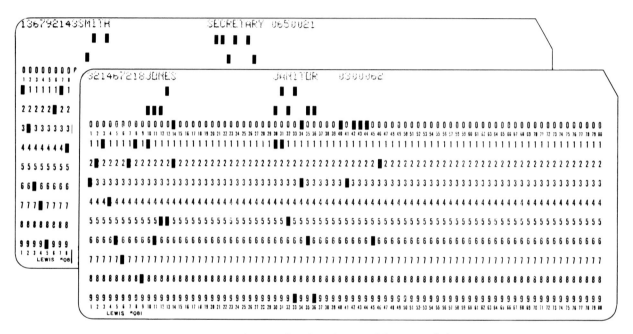

Figure 4.1. Sample record layout and data

value $6.50 would be entered as 0650. Note that the decimal point is not punched or typed on the record; the processing program assumes the decimal point to be between columns 41 and 42.

Associated with each record is some identifying number or name, sometimes referred to as the *key field*. For example, a personnel file might use the employee's social security number or his name as a key field. An accounts receivable file probably would use as the key field an account number assigned to the customer. Key fields are used in retrieving or updating records in a particular file.

A common means of organizing a file is to store the records in such a way that key fields are in *sequential* order.[1] Either *ascending*

1. Other ways of organizing files include random, and index-sequential organization but are beyond the scope of this text.

or *descending* sequence may be used. With ascending sequence each succeeding record within a file has a key field value larger than the preceding record. When descending sequence is used each record has a key with a lesser value than its predecessor. Of the two methods, ascending sequence is more commonly used. Figure 4.2 shows a simple payroll file. The key field is the social security number.

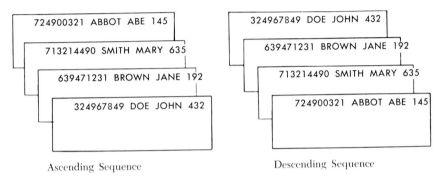

724900321 ABBOT ABE 145
713214490 SMITH MARY 635
639471231 BROWN JANE 192
324967849 DOE JOHN 432

324967849 DOE JOHN 432
639471231 BROWN JANE 192
713214490 SMITH MARY 635
724900321 ABBOT ABE 145

Ascending Sequence   Descending Sequence

Figure 4.2. Ascending and descending file organization

In order to process a file, a computing system requires a device capable of reading information from and/or writing information onto an appropriate recording medium. Such devices as card readers, paper tape readers, tape drives, disk drives, and drums are commonly attached to the CPU for this purpose.

Files in which records are organized sequentially are most often *accessed* in sequential fashion, that is, the records of a file are made available to the processing program, one record at a time, in the order they were stored. This method is usually required for files stored on cards, paper tape, and magnetic tape and optional for files stored on magnetic disk, drums,[2] and diskettes (microcomputers). The logic required to process sequential files is essentially independent of the device on which the file is stored.

Regardless of the content of a file and the medium used for its storage there are common tasks that must be performed to create and maintain that file within an information processing system. Among these are:

2. Disk, drum, and other related devices are sometimes referred to as "direct access storage devices." Files may be stored in such a way that individual records within the file may be accessed without processing other records in the file.

1. creating the file in proper sequence
2. merging the contents of two files
3. adding records to the file
4. deleting records from the file
5. changing (updating) individual records within the file
6. generating detailed and/or summary reports

The sequence of operations to be performed and the flow of information among files and processing programs is often described by means of a system flowchart. In this chapter we will briefly examine these flowcharts and subsequently treat the logic required to create and maintain sequential files.

4-2 SYSTEM FLOWCHARTING

Flowcharting is often used not only to represent the logic of an algorithm but also to represent the sequence of processing operations and the flow of information within an information processing system. The symbols shown in Figure 4.3 are often used in such *system* flowcharts. Each symbol used is identified with a report name, file name, or program name as shown in Figure 4.4. When a file is required as input into a processing program a flowline is drawn connecting the file symbol to the processing block. When a file or report is produced as output from a program a flowline is drawn from the processing block to the file or report block. For example, consider Figure 4.5. The program "UPDATE" requires as input a tape file, "MASTER FILE," and a card file, "TRANSACTIONS," and produces a report, "TRANSACTION LISTING," and a disk file, "UPDATED MASTER FILE." Note that the arrow head on the flow line indicates whether the file is used as input or produced as output.

An information processing system often involves a sequence of processing operations. A system flowchart such as shown in Figure 4.6 is especially useful in visualizing the relationships among the various files and programs. For example, in Figure 4.6 the "UPDATE" program must be run to produce the "UPDATED CUSTOMER MASTER FILE" before the "BILLING" program can be used to produce the customer statements. System flowcharts serve as an aid to the systems analyst who designs the entire information processing system, to the programmer who writes and tests individual programs within the

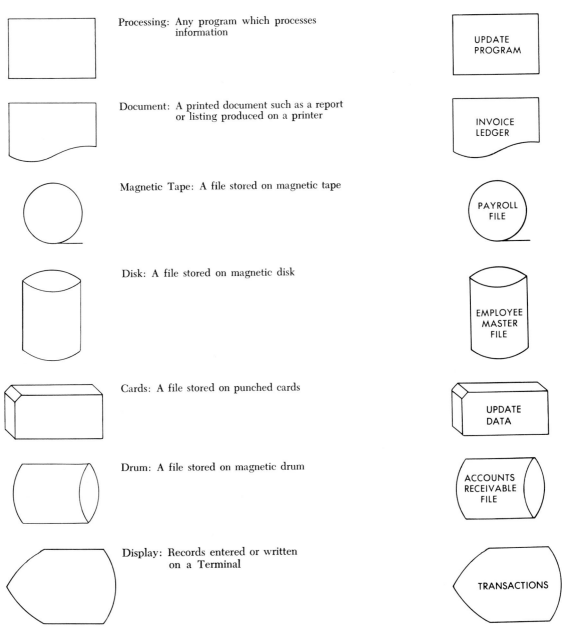

Processing: Any program which processes information

UPDATE PROGRAM

Document: A printed document such as a report or listing produced on a printer

INVOICE LEDGER

Magnetic Tape: A file stored on magnetic tape

PAYROLL FILE

Disk: A file stored on magnetic disk

EMPLOYEE MASTER FILE

Cards: A file stored on punched cards

UPDATE DATA

Drum: A file stored on magnetic drum

ACCOUNTS RECEIVABLE FILE

Display: Records entered or written on a Terminal

TRANSACTIONS

Figure 4.3. System flowchart symbols

Figure 4.4. Examples of system flowchart symbols

system, and to the operator who must run the programs in proper sequence and with proper data to ensure the correct functioning of the
system.

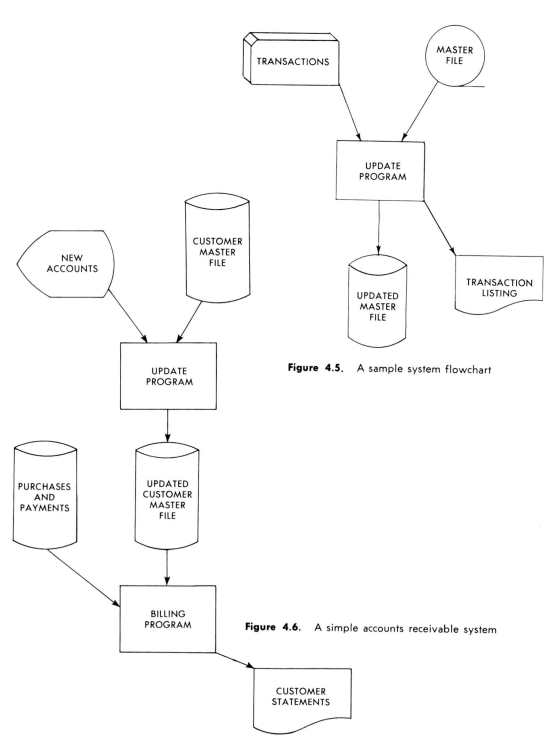

Figure 4.5. A sample system flowchart

Figure 4.6. A simple accounts receivable system

At the inception of any information processing system a file must be created. Data must be gathered and prepared in machine readable form. Terminals are a common form used for initial entry of data into a system.[3] The created file must have records with key fields in sequential order. As noted earlier, ascending sequence is most commonly used. If the initial data file has been correctly sorted into ascending sequence, the file creation program could simply read a record, move it to an output area, and write this record on the output file.

Reading an entire record as opposed to reading fields from one record can be described by an input block such as

/ READ A RECORD / where the record can consist of one or more
/ FROM FILE A / fields as in Figure 4.1

When processing multiple input/output files, it is important to mention the name of the file within the input/output block to avoid confusion. Writing an entire record can be accomplished by means of a statement such as:

/ WRITE A RECORD / A complete record with one or more
/ ONTO FILE B / fields is written out

In the case of file processing algorithms, such as file creation, the format of an input record will generally not be the same as the master file format for that record. For example, fields in the master record may be extended in length; the order of the fields may change; fields may exist in the master file that are not present in the input record. Before an input record can be recorded on an output file, it must usually be transformed to conform to the master file format. This can be shown in the flowchart by the processing block

| MOVE INPUT RECORD TO NEW FILE OUTPUT | Implied in this block is a series of moves, the specifics of which would depend on the files being processed.

The correct functioning of a file processing system depends on files which have records in sequential order. It is, therefore, a good practice to perform a *sequence* check in the file creation program, that is, verify that records are in proper sequence. Records from the initial

3. Other common forms of data entry are key to tape, and key to disk. Optical character readers and magnetic ink character readers are also used for specialized applications.

file which are not in proper sequence should not be included in the output file. A list of records which are out of sequence could be produced for later addition to the file. A system flowchart for file creation is shown in Figure 4.7. A program flowchart for the file creation program is shown in Figure 4.8. The assumption is made that all records will have a key field value greater than zero. The variable HOLD is initialized to zero and, after the first record is processed, is used to hold the value of the key field previously written onto the output file. By comparing the key field on each input record to the value in HOLD records which are out of sequence can be determined.

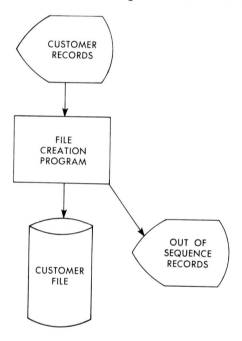

Figure 4.7. File creation

EXERCISES 4-1

1. A new tape file is to be created from records found in three sequential input files A,B,C, the records of which all have the same layout format and where each key field is 4 digits long.
   File A is a card file with records in which the key fields ranges from 001 to 500.

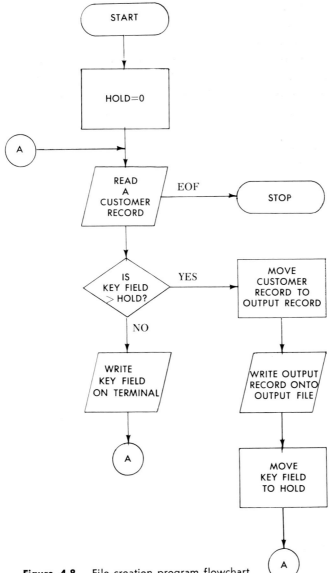

Figure 4.8.   File creation program flowchart

File B is a tape file consisting of at least 50 records with key fields ranging from 600 to 700.

File C is a disk file with key fields 750 to 999.

Draw a system flowchart to create a disk consisting of records from all three input files. Draw a program flowchart for the program.

2. Draw a system flowchart to describe the following information processing system: Widget Production Inc. maintains a parts inventory control system. The master file is stored on disk; there is one record in the file for each part the company uses in manufacturing widgets. The system is used to make sure enough of each type of part is available at all times so that production can continue without interruption. An on-line system is available to collect data regarding new types of parts that may be needed as different models of widgets are designed and produced. The system also collects data regarding parts used and new parts received from supplies. Periodically these transactions are sorted into appropriate order and a program is run to update the master file. At this time a program is run to order more parts from suppliers if the number on hand falls below a specified level.

3. Trace the execution of the algorithm shown in Figure 4.8 for the data file in Figure 4.9. The key field is the first nine digit field.

4. Is it valid to have two records in a system with the same value in the key field? How will such records be treated by the algorithm in Figure 4.8? How could the algorithm be modified to allow records having the same value in the key field to enter the file?

4-4 FILE UPDATING

Once a file has been created procedures must be devised to maintain the file so that it reflects changes which take place. New employees are hired, others retire or are fired; some employees receive raises in salaries, others change their marital status or number of exemptions. Payroll files must be updated before pay checks can be written. A simple form of updating is merging two files together to form one sequential file as shown in Figure 4.10.

Merging consists of taking records from two (or more) files and creating a single new file still in sequential order. Consider, for example, the two old files A and B and the new file shown in Figure 4.11. A program flowchart to merge two sequential files is shown in Figure 4.12. Initially a record is read from each of the two files. The record with the smaller key field value is written onto the new file. Another record is then read from the file whose record was just written on the new file. The key field from this new record is compared to the other

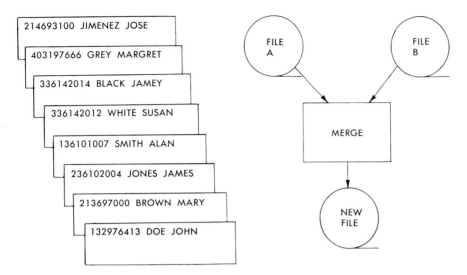

214693100 JIMENEZ JOSE

403197666 GREY MARGRET

336142014 BLACK JAMEY

336142012 WHITE SUSAN

136101007 SMITH ALAN

236102004 JONES JAMES

213697000 BROWN MARY

132976413 DOE JOHN

Figure 4.9.

FILE A

FILE B

MERGE

NEW FILE

Figure 4.10....System flowchart for merging example

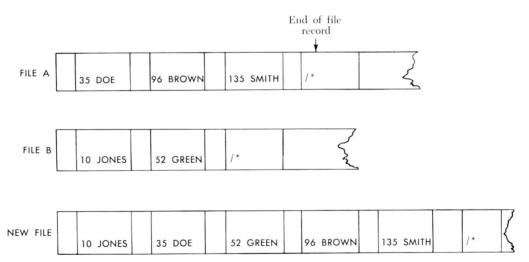

End of file record

FILE A | 35 DOE | 96 BROWN | 135 SMITH | /* |

FILE B | 10 JONES | 52 GREEN | /* |

NEW FILE | 10 JONES | 35 DOE | 52 GREEN | 96 BROWN | 135 SMITH | /* |

Figure 4.11.   Merging example

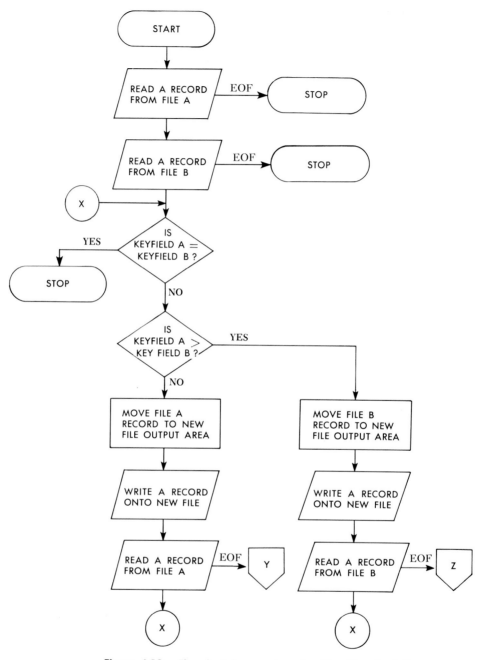

Figure 4.12. Flowchart for merging algorithm (Part 1 of 2)

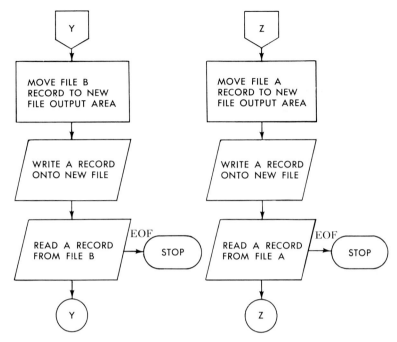

Figure 4.12.   Flowchart for merging algorithm (Part 2 of 2)

existing key field and the procedure is repeated until all of the records from one of the files is exhausted. The remaining contents of the other file can simply be copied onto the new file.

In many instances it is desirable to process additions and deletions to a file concurrently. A program will then process these changes to produce an updated master file in which new records have been added and old records have been deleted or modified as appropriate. The new file will be produced on an entirely new reel of tape or area of disk. The old file and the changes will be saved in case errors have occurred in the update process or the new file is inadvertently destroyed. In most systems at least two (and usually more) generations of files and associated changes are preserved for this purpose.[4]

4. When files are stored on disk or drums it is technically possible to change existing records (i.e., perform an update) without creating an entirely new file. When this is desired, random or index sequential file organization is usually used.

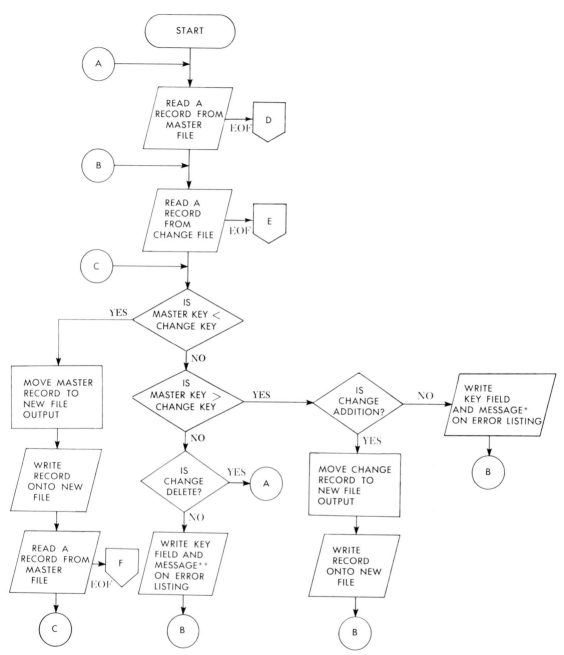

*"KEY FIELD NOT FOUND" An attempt has been made to delete a record not present in the file.

**"DUPLICATED KEY FIELD" An attempt has been made to add a record with the same key field value as a record already present in the file.

Figure 4.13.   Adding and deleting records (Part 1 of 2)

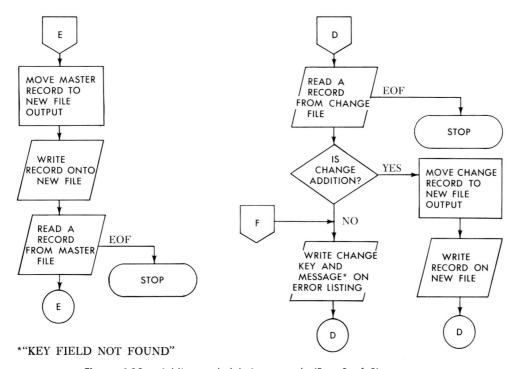

*"KEY FIELD NOT FOUND"

Figure 4.13. Adding and deleting records (Part 2 of 2)

A sample system flowchart for the file update procedure is shown in Figure 4.5. A flowchart for a program which will add or delete records from a file is shown in Figure 4.13. It is assumed that the change records are coded with a key indicating whether or not a record is to be added or deleted and are in the same sequence as the master file. Note that several types of errors may be encountered in the update process. An attempt may be made to delete a record not in the file. A record may enter the system with the same key field value as a preexisting record. Such errors should be noted with an appropriate error message.

When changes are to be made to existing records in a file, the change records must be coded with the type of change to be made. Is it the name, address, marital status, or pay rate that is to be changed? A general flowchart for a program to change existing records in a file is shown in Figure 4.14. Note that all changes in fields are made in the new file output area. The record is written onto the new file only after a new change record has been read and it has been determined

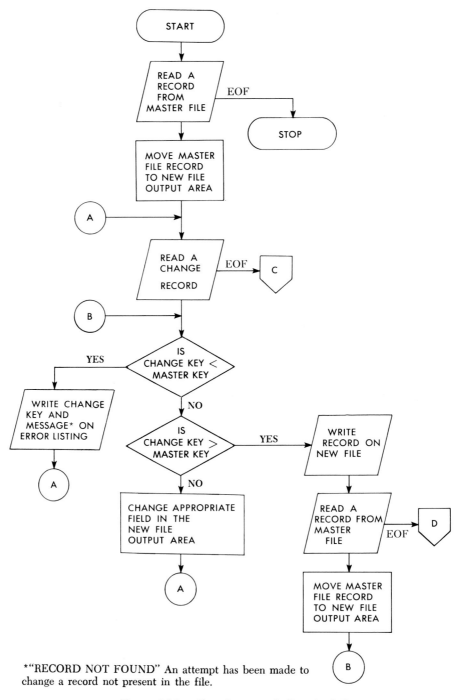

*"RECORD NOT FOUND" An attempt has been made to change a record not present in the file.

Figure 4.14. Changing records (Part 1 of 2)

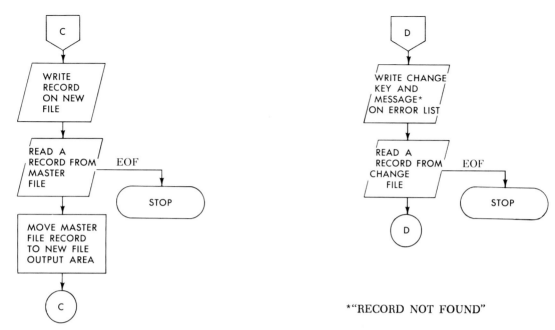

Figure 4.14.   Changing records (Part 2 of 2)

that the change does not pertain to the record currently stored in the output area. In this way many changes may be made to one record. Again error conditions may arise and should be noted.

EXERCISES 4-2

1. Given the two files A and B shown in Figure 4.11 trace the execution of the algorithm shown in Figure 4.12.
2. In Figure 4.12 suppose there are no records in either FILE A or FILE B. How does the algorithm handle this case? Is this desirable? What else might be done?
3. In Figure 4.12 suppose a record in FILE A and a record in FILE B have the same value in the key field. How does the algorithm handle this case? Is this desirable? What else might be done?
4. Draw a flowchart for an algorithm to merge two sequential files. The new file should contain *all* the records from the original files and be in sequential order.
5. Given the master file shown in Figure 4.15, display the output file produced as a result of executing the algorithm in Figure 4.13 with the change file as in Figure 4.16.

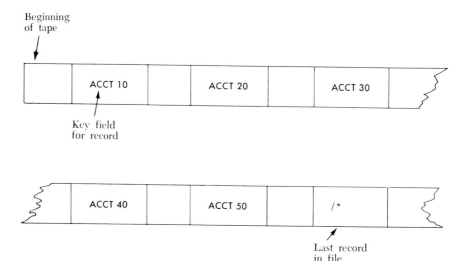

Figure 4.15.

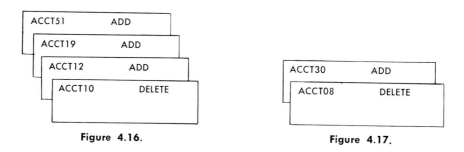

Figure 4.16.   Figure 4.17.

6. Same exercise as 5 with change file as in Figure 4.17.
7. Given the master file shown in Figure 4.18 and the change file in Figure 4.19, display the output file produced as a result of the algorithm in Figure 4.14.
8. Same exercise as 7 with change file as in Figure 4.20.
9. Same exercise as 7 with change file as in Fig. 4.21.
10. In view of question 9 is it wise to modify the flowchart in Figure 4.15 to include a sequence check for the change file?
11. Given a master file, draw a flowchart to produce a new master file as a result of adding, deleting, and making changes to records in the original master file.

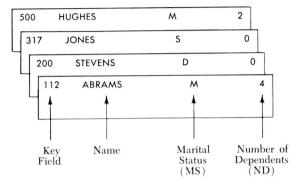

Figure 4.18.

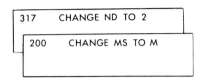

Figure 4.19.

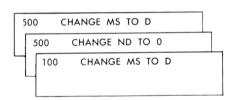

Figure 4.20.

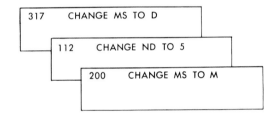

Figure 4.21.

12. Every day the Podunk College Library receives newly purchased books, returned books, and books given to the college as gifts. Books are removed from the shelves because of mutilation or theft, because they are out of date, and because they are borrowed from the library by patrons. Records of each of these transactions are entered at a terminal and stored in separate disk files.

FILE IN contains one record for each book returned by a borrower.

FILE OUT contains one record for each book borrowed.

The record layout for FILE IN and FILE OUT is

Book #	In/Out Code	Date	Faculty/ Student Code	Address

FILE A contains one record for each new book purchased and received.

FILE B contains one record for each book received as a gift. The record layout for FILE A and FILE B is

Book #	Edition/ Year	Gift/ Purchase Code	Author	Reserve Status

FILE X contains one record for each change made by a librarian in the status of a book. Books may be deleted from the file, or the reserve/available status may be changed. The record layout for FILE X is

Book #	Deletion Code	Reserve Status

All of these files ultimately are used to update the MASTER FILE which contains one record for each book in the library. A partial record layout for MASTER FILE is

Book #	Edition/ Year	Gift/ Purchased Code	Reserve Status	Faculty/ Student	In/ Out	Author ...

(In a real system, other fields would undoubtedly be included in a master file record.)

Draw a system flowchart and individual program flowcharts to accomplish each of the following:

1. Books from FILE A and FILE B must be added to MASTER FILE daily.
2. MASTER FILE must be updated from FILE IN and FILE OUT daily.
3. MASTER FILE must be updated from FILE X daily.
4. Books with edition/year less than edition/year of newly purchased or given books (FILE A and FILE B) must be taken out of circulation daily. A printout and a disk file in FILE X format must be generated. This file serves as additional input into FILE X.
5. Once a month a list of all books in the library received as gifts must be provided. This list must be derived from the master file.

REPORT WRITING 4-5

The data stored in records on a file is generally of little value until it is summarized, that is, reduced to usable form, and made available to persons who can use the information thus produced. Such common summary statistics as grand totals, averages, and counts have been considered in preceding sections. In many instances, however, more refined summaries may be required.

For example, suppose we have a file in which each record contains a salesman number, a date of sale, and an amount of sale. We will assume that the records have been sorted into order by salesman number; all records for a given salesman will thus occur in a group. Consider the sample file shown in Figure 4.22. Note that the number of records for a given salesman is variable; for example salesman 004 made three sales, salesman 006 made one sale, etc. A report is desired which will summarize not only the total amount of sales, but also the total sales for each salesman. Such a report for the sample data in Figure 4.22 is shown in Figure 4.23. The totals for each salesman are called *intermediate* totals or *minor* totals.

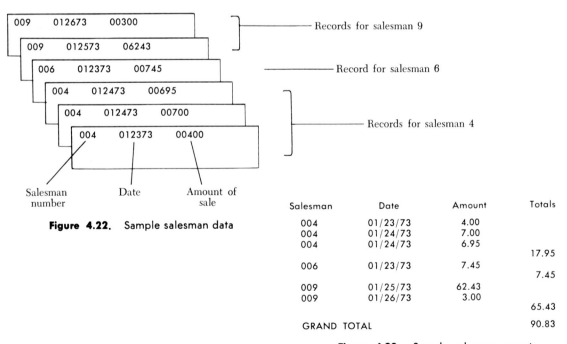

Figure 4.22. Sample salesman data

Salesman	Date	Amount	Totals
004	01/23/73	4.00	
004	01/24/73	7.00	
004	01/24/73	6.95	
			17.95
006	01/23/73	7.45	
			7.45
009	01/25/73	62.43	
009	01/26/73	3.00	
			65.43
GRAND TOTAL			90.83

Figure 4.23. Sample salesman report

A program flow chart to produce this report is shown in Figure 4.24. The basisc idea in accumulating minor totals is to search for a change in salesman number as the records are read. If a record pertains to the same salesman as the preceding record, the amount of sale will be added to the current minor total. If the record pertains to a new salesman a *control break* has been found. The previous minor total is written out at this point and a new minor total is initialized for the new group of records.

The decision as to whether a record belongs to a new group or the current group is made by storing the salesman number of the current group in some location (HOLD in the flowchart in Figure 4.24) and comparing the salesman number to that location each time a record is read. If the two locations have the same value the record belongs to the current group; if they are unequal a control break has been determined.

How should the value of HOLD be initialized to avoid a control break when the first record is read? There is no way for the program to know ahead of time the first salesman's number. This problem can be handled by reading the first salesman's number and then initializing HOLD. In the case of Figure 4.24 the first READ statement and the replacement statement which follows is used to initialize HOLD to the first salesman's number which will be encountered in the input file. A meaningful comparison between HOLD and SNO can then take place.

EXERCISES 4-3

1. Suppose records are out of sequence when the algorithm of Figure 4.24 is executed. How will this affect the minor totals? the grand total?

2. Construct a program flowchart to produce a report showing the grade point average and the total of credits earned by each student and the overall grade point average for all students. The description of input records is as follows:

> student number
> course name
> course grade (4=A, 3=B, 2=C, 1=D, 0=F)
> credit hours

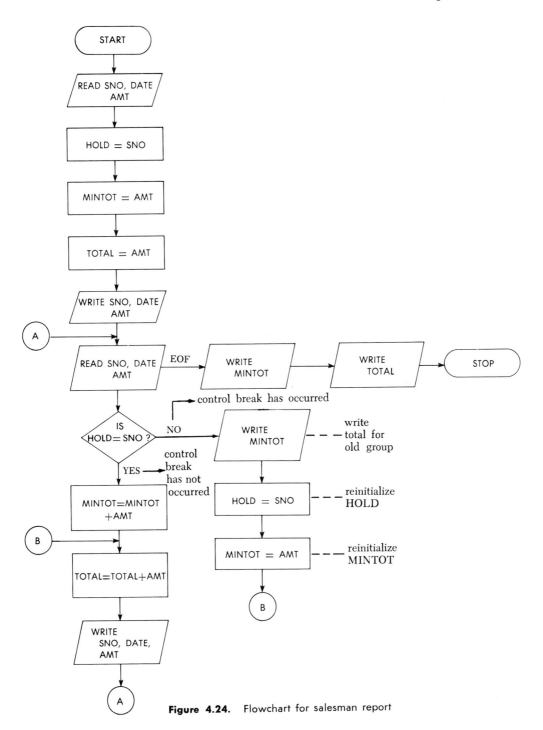

Figure 4.24.   Flowchart for salesman report

There is one record for each student enrolled in each course. Assume that the data is sorted into sequence by student number.

4-6 STRUCTURED PROGRAMMING

As the cost of hardware has decreased and the cost of programming has increased, there has been a shift to *program modularity*. This new programming approach is called *structured programming*. The three basic building blocks for structured programming modules are shown in Figure 4.25. Any programming task which must be performed can be carried out using modules of these three types.

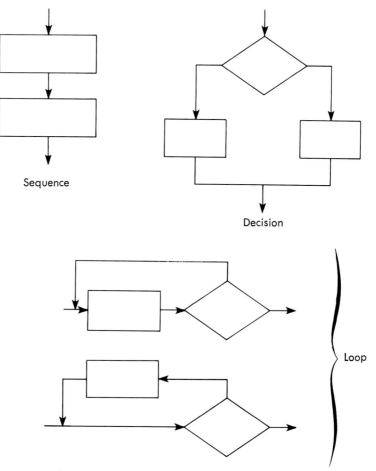

Figure 4.25. Structured Programming Modules

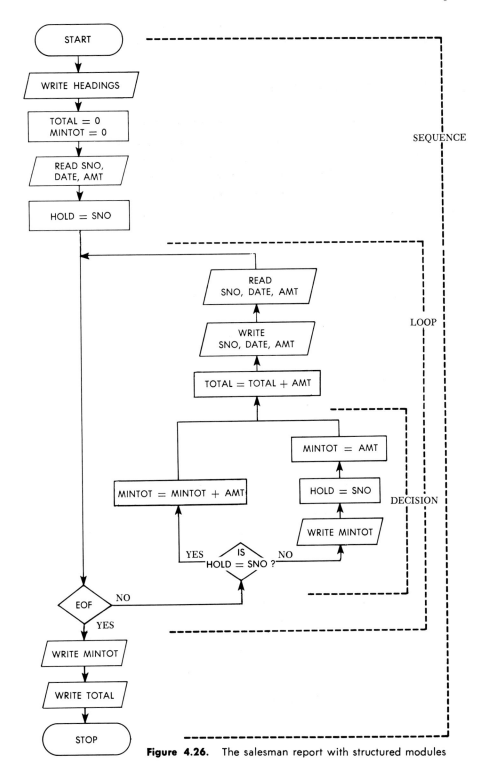

Figure 4.26. The salesman report with structured modules

The flowchart for the salesman report (Figure 4.24) is redone in structured format in Figure 4.26. Figure 4.27 breaks down the complete program of Figure 4.26 into smaller modules.

The basic program structure is a sequence module shown on the left of Figure 4.27. It is a set of tasks which are executed sequentially. One task—Processing records until end of file—is implemented as a loop type module shown to the right of the basic sequence module. The loop module in turn requires a module to Process a record and read the next record. This module is basically a sequence type module, however one task—Check for control break in saleman number and take appropriate action—is implemented as a decision module which is shown as the right most module in Figure 4.27.

In structured programming, any program module may be thought of as a complete functioning whole. Any module may be made up of smaller modules and may be a part of a larger module. A basic advantage to structured programming is the ability to concentrate on the achievement of the objectives for each program module. Any task which is complex may be described in general terms and becomes another smaller module in the program. This technique is formally called *top down programming*. The use of top down design and the restriction of using only the three types of module structures form the basic ideas of structured programming.

EXERCISES 4-4

1. Using only the three structured programming modules, draw a flowchart for a program that could be used by a company to determine the weekly payroll. The input data is one record per employee with the following fields: employee name, employee number, hourly rate of pay, and number of hours worked. Have your solution list the employee name, employee number, hourly rate of pay, number of hours worked, and gross pay. Remember to pay time and one half for all hours over 40.

2. Add a summary total of all gross pay to exercise 1.

3. Using only the three structured programming modules, draw a flowchart that could be used to update an inventory file. Provide for sales, sales returns, purchases, purchases returns, and lost items.

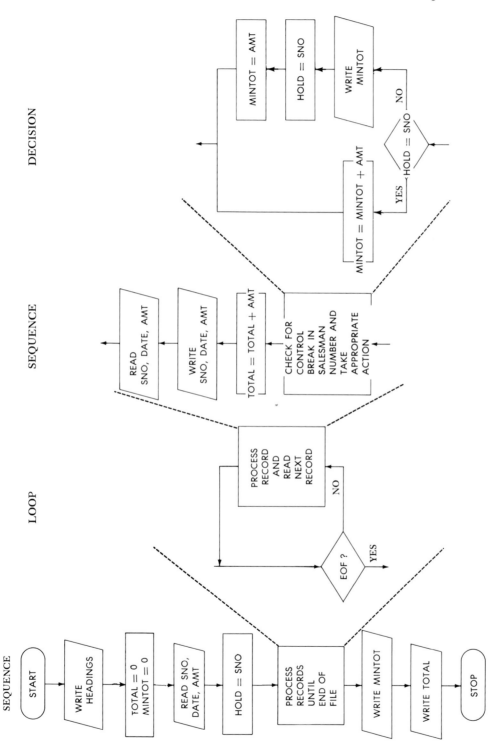

Figure 4.27. Salesman report broken into smaller modules

5 Arrays

An array is a collection of contiguous storage locations sometimes called *elements* which may be uniquely identified by specifying (1) the name of the given array (common to all elements of that array), and (2) the position of that element within the array.[1] The most powerful feature of the array concept is that the array elements may be indexed. This indexing mechanism can be translated mathematically by subscripting variables. Consider, for example, the array A pictured in Figure 5.1.

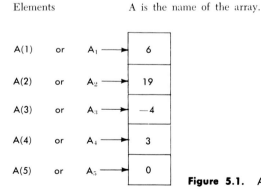

Elements   A is the name of the array.

A(1)	or	A_1	→	6
A(2)	or	A_2	→	19
A(3)	or	A_3	→	−4
A(4)	or	A_4	→	3
A(5)	or	A_5	→	0

Figure 5.1. An array

1. The position of the element is designated by the use of one or more subscripts. Only arrays with one subscript will be treated in this text.

The array in Figure 5.1 consists of five elements A_1, A_2, A_3, A_4, A_5 (where A_1 is read as A sub 1, etc.) The value of A_1 is 6, that of A_2 is 19, etc. Since it is not possible to represent subscripts in the above form on most input devices, an alternate convention is to write $A(1)$, $A(2)$, $A(3)$, . . . etc. Both subscript representations are widely accepted in flowcharting though the latter is more readily adaptable to program coding in most languages.

From (2) of the definition of an array it follows that subscripts must be integers greater than zero. A subscript may be a constant or a variable; for example, if K equals 4, $A(K)$ refers to the fourth element of the array A. If, K equals zero, $A(K)$ is invalid as there is no *0th* element of A.[2]

Suppose there are five numbers on a record and it is desired to compute their sum. One approach to the problem is to read these numbers into five separate variables and compute their sum as shown in Figure 5.2. An alternate approach is to store the five values in an array. The sum of the elements may then be computed directly (See Flowchart A, Figure 5.3) or accumulated within a loop (See Flowchart B, Figure 5.3). In the latter case, the first time through the loop SUM = SUM + $A(K)$ = $A(1)$ since SUM = 0 and K = 1; the second time through the loop SUM = SUM + $A(K)$ = $A(1)$ + $A(2)$ since SUM = $A(1)$ and K = 2; partial sums are thus formed until the total sum has been accumulated. In this simple example, there seems to be nothing gained by the use of an array over unsubscripted variables. The advantage of arrays is that this type of data structure lends itself very conveniently to loop processing and thereby to accumulation type processes. Suppose, for instance, that 1,000 numbers read from a file were to be added; providing variable names for all these values and expressing the addition statement would be, to say the least, unrealistic. Storing these values in an array and running the array through a loop to accumulate the sum would be markedly more efficient.

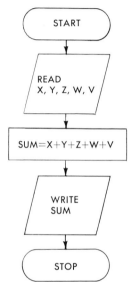

Figure 5.2. Sum of five values without an array

5-2 LOADING AN ARRAY

Loading an array with data contained on a single record, has been accomplished so far by specifying all elements of the array in the list of variables of the READ statements as in Figure 5.3. This

2. Some computer language will permit a reference to A (0).

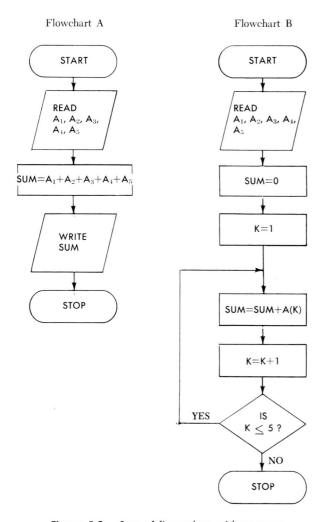

Figure 5.3.   Sum of five values with an array

Figure 5.4.   Alternate form for specifying a list of array elements

method becomes quite tedious when numerous values are to be read into the array. A convenient and more efficient notation to achieve the same result is shown in Figure 5.4. The input block is interpreted as read A(I) as I takes on the successive integer values 1, 2, 3, . . . 5 meaning read A(1), A(2), . . . A(5) in that order.

Loading an array from more than one record where each record contains only one value can be accomplished by placing the input

block, READ A(I) in a loop where I is incremented by 1 each time through the loop. Repeated execution of the READ instruction causes new records to be read while assigning the corresponding record values to different array elements each time. Figure 5.5 illustrates loading of an array from five different records.

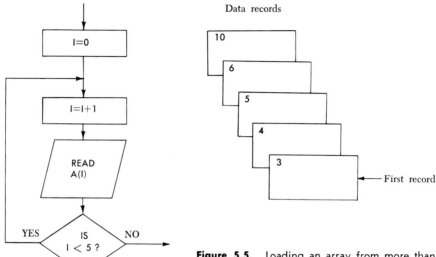

Figure 5.5.   Loading an array from more than one record

5-3 A PRACTICAL EXAMPLE

The desirability of the array feature can be illustrated by the following example: A file contains 35 grades, one grade per record. It is desired to determine the number of grades below the class average. The average can be computed without storing all the grades in memory as was done in Section 3-4 by reading one grade at a time and accumulating the total as the grades are read. However, if this is the approach taken, it becomes impossible to determine the number of grades below the average since reference to each grade is no longer possible, the latter having been destroyed by each successive READ statement. To compare each grade with the average, it is essential that all of the grades must be present in memory at the same time. Figure 5.6 illustrates a solution to this problem.

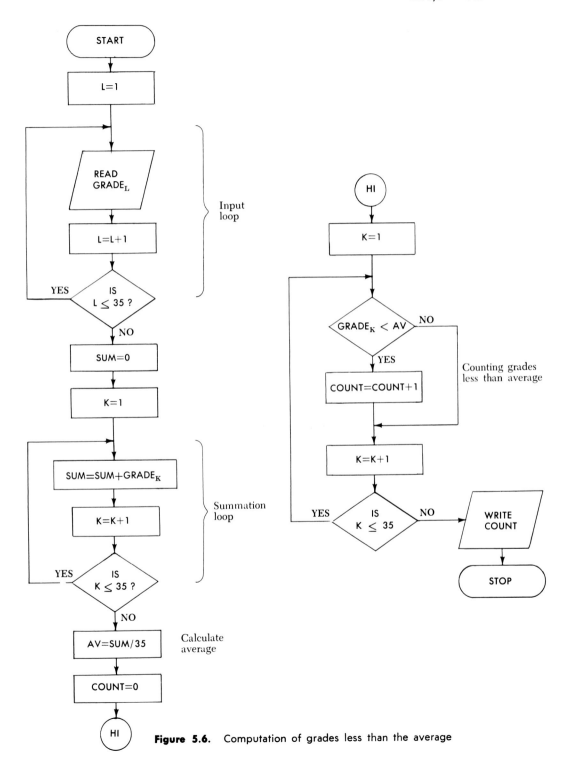

Figure 5.6. Computation of grades less than the average

In many instances, it is necessary to initialize the contents of an array to some particular value before the array is processed. Figure 5.7 illustrates such an initialization which results in setting all array elements to zero.

5-4 TABLE LOOKUP

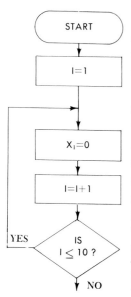

The table lookup process is a fast and efficient method to access data in a table. Access is *direct* in the sense that no searching of the table need take place. This process can be understood by considering the following example:

The shipping rate per pound of merchandise to specific postal zones is shown in Figure 5.8. For example, COST (1) = \$.10 means the cost to send 1 pound of merchandise to postal zone 1 is \$.10. It is assumed that the shipping rate is directly proportional to the weight of the merchandise, i.e., if it costs \$.10 to send one pound of merchandise to postal zone 1, it will cost \$.20 to send 2 pounds of merchandise, etc. Each record in a file contains two fields; the first is a postal zone number Z, and the second is the weight of a package WEIGHT. The problem is to determine the cost of shipment of each package to its destination. We will assume that the table of rates has already been loaded into the COST array by an initialization process (See Figure 5.9). If the zone Z read from a record has value 3, then the corresponding shipping rate is COST(Z) = COST(3) = \$.14 so that the total costs of shipping that package to zone Z can be computed as TOTCOST = COST(Z) ° WEIGHT for the value Z and value

Figure 5.7. Array initialization

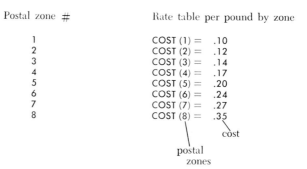

Postal zone #	Rate table per pound by zone
1	COST (1) = .10
2	COST (2) = .12
3	COST (3) = .14
4	COST (4) = .17
5	COST (5) = .20
6	COST (6) = .24
7	COST (7) = .27
8	COST (8) = .35

cost

postal zones

Figure 5.8. Rate table

WEIGHT read from the same data record. Figure 5.9 shows the complete flowchart to solve the above problem.

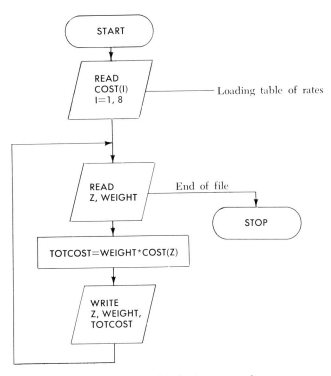

Figure 5.9.   Table lookup example

FREQUENCY DISTRIBUTION 5-5

A file contains grades ranging in value from 1 to 100. It is desired to calculate a frequency distribution for those grades, i.e., count the number of occurrences of each grade. For instance, if the grade 78 appears 5 times in the input file, then the frequency of the grade 78 is five. Since there are 100 possible grades, one hundred counters will be required to record all of the occurrences. It is, of course, possible that some of the values will never appear in the file. Partly for that reason it becomes necessary to initialize all counters to zero before the counting process starts (see Figure 5.10). As each data rec-

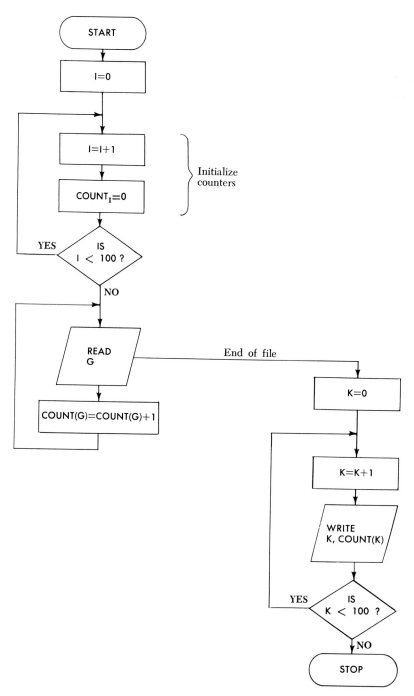

Figure 5.10.   A frequency distribution

ord is read, the value of the variable G is used to designate which of the counters should be incremented. Thus, if G has the value 65, the block $\boxed{\text{COUNT}(G) = \text{COUNT}(G) + 1}$ will cause the value of COUNT(65) (i.e., the counter used to record occurrences of the grade sixty-five) to be incremented by one. Note the method used to write out the elements of the array count (see Figure 5.10).

SORTING 5-6

Often it is desirable to sort data into ascending or descending sequence. Various algorithms exist for sorting. If the amount of data is too large to be stored in memory at one time, sort algorithms which make use of temporary holding files to store data during intermediate stages of the sort process are generally used. These algorithms are rather complex and beyond the scope of this text. If, on the other hand, the amount of data is sufficiently small to be stored in an array in memory, a very interesting algorithm exists for sorting.

Consider the data in the beginning array of Figure 5.11 which is to be sorted into ascending order. A procedure that can be used is to shift the largest value in the array to the last position, then shift the next largest value to the next to last position and so on.

The shifting process is accomplished by comparing consecutive positions within the array and interchanging the elements when necessary. For example, A(1) and A(2) are compared. For the data in

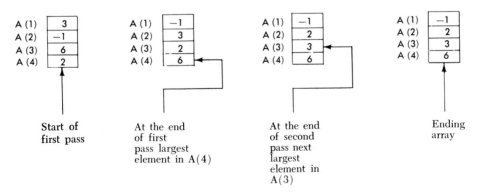

Figure 5.11.

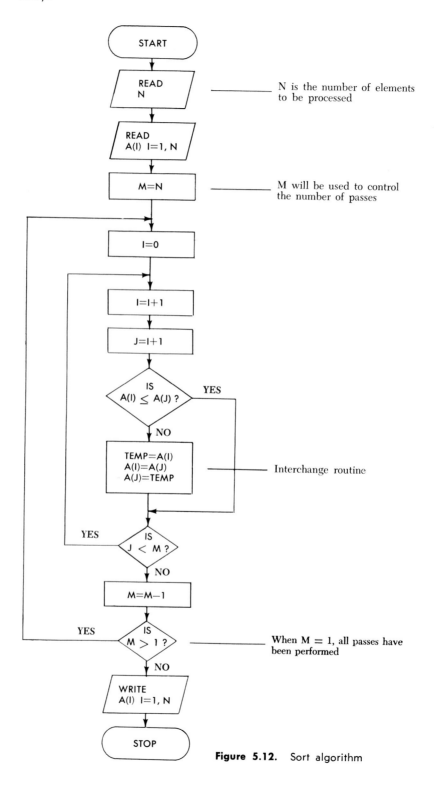

Figure 5.12. Sort algorithm

Figure 5.11, A(1) > A(2) since 3 > −1 and the elements must be interchanged. After performing the interchange the array will appear as:

A(1)	−1
A(2)	3
A(3)	6
A(4)	2

Then A(2) and A(3) are compared. A(2) < A(3) since 3 < 6 and therefore the interchange is not performed. Finally, A(3) and A(4) are compared and the interchange is performed. The array now has its largest value in the last position. The process must now be repeated, i.e., a second pass must be made through the array, but this time the last element A(4) need not be considered. At the end of the second pass the data in this example is in sorted order. This, however, need not always be the case. All that can be said with certainty is that the lasts two positions are in ascending sequence. In general, a third pass must be performed to sort the remaining elements into proper sequence. In theory, if an array has *n* elements, this algorithm will require at most *n*-1 passes to sort the data completely. A flowchart for this algorithm is shown in Figure 5.12. Note the interchange routine in this algorithm. A location, TEMP, is required so that the value in A(I) is not destroyed by the replacement A(I) = A(J).

EXERCISES 5-1

1. Tabulate the values assumed by the variables in Flowchart B Figure 5.3 for the array shown in Figure 5.1.
2. Is it possible to combine the input loop and summation loop in the flowchart in Figure 5.6? Is it also possible to combine the counting loop?
3. Draw a flowchart to fill successive elements of an array with values 1, 3, 5, 7, . . . 21.
4. Attempt the construction of a flowchart for the problem of Section 5-5 without using an array. Is it possible? Is it practical?
5. Three records each containing thirty numbers are to be read. Draw a flowchart to store these ninety numbers in an array XRAY and compute their sum and product.

6. Suppose we wish to include the possibility of a grade of zero in the problem of Section 5-5. How can the problem be handled? (Remember that the value zero may not be used as a subscript.)

7. What output will be produced by the flowchart in Figure 5.9 for the file in Figure 5.13?

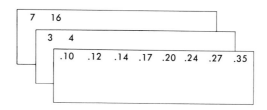

Figure 5.13.   Sample data file

8. Draw a flowchart to input two vectors[2] A and B each with five components $A_1, A_2 \ldots A_5, B_1, B_2 \ldots B_5$ and compute the inner product P

$$P = A_1 * B_1 + A_2 * B_2 + \ldots + A_5 * B_5$$

9. Draw a flowchart to compute the sum of the squares of the elements of an array with fifty elements.

10. Draw a flowchart to read an input file containing one grade per record and output each grade and its deviation from the average. The deviation is found by subtracting the average from the grade.

11. An array contains n elements $A_1, A_2, \ldots A_n$. Draw a flowchart to assign values to array B such that

$$B_1 = A_n, B_2 = A_{n-1}, \ldots B_n = A_1$$

12. Draw a flowchart to determine the location of the lowest grade of an array containing 100 grades, i.e., not its value but its position in the array.

13. Trace the execution of the algorithm in Figure 5.12 for the data file shown in Figure 5.14.

2. An array of single dimension may be used to represent a vector.

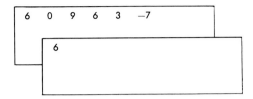

Figure 5.14. Sample data

14. Flowchart a program to:
    a. Read the sales figures for the last twenty-four months into an array.
    b. Compute the average monthly sales for the twenty-four-month period.
    c. List the months in which the sales were 15 percent greater than the average.
    d. List the months in which the sales were 5 percent less than the average.

5-7 SUBPROGRAMS

Frequently parts of a program will be identical to parts of a previously written program. Also, identical steps may be repeated in several parts of a program. A *subprogram* is a procedure designed for a particular task; it may be used to save time and effort in either of these cases. In the first case, subprograms eliminate the need to re-flowchart (and re-code) a previously designed procedure. In the second case, a procedure may be designed once and used at various points of a program; thereby eliminating needless re-coding of similar logic.

A *main program* is used to call for the execution of a subprogram. A subprogram is referenced in a main program by the use of the *pre-defined* process block which has the general form:

```
‖ subprogram ‖
‖    name    ‖
```

For example, the sort procedure developed in section 5-6 can be written as a subprogram as shown in Figure 5.15. In flowcharting a subprogram, the termination block 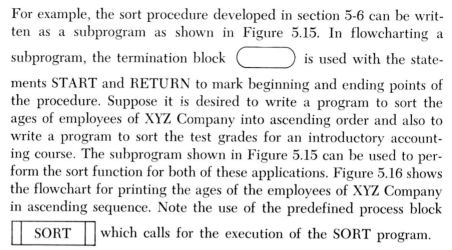 is used with the statements START and RETURN to mark beginning and ending points of the procedure. Suppose it is desired to write a program to sort the ages of employees of XYZ Company into ascending order and also to write a program to sort the test grades for an introductory accounting course. The subprogram shown in Figure 5.15 can be used to perform the sort function for both of these applications. Figure 5.16 shows the flowchart for printing the ages of the employees of XYZ Company in ascending sequence. Note the use of the predefined process block

| SORT | which calls for the execution of the SORT program.

Figure 5.17 shows the flowchart for printing the grades for the introductory accounting course. Although the main programs for each of these applications differ in some respects, the procedure required for sorting the data is the same and hence the subprogram SORT can be used to good advantage in both applications.

EXERCISES 5-2

1. A data file consists of student names and one test grade per student. There may be up to 100 records of input. Load a table with this data, compute the average grade and produce a list of the table and the deviation of each grade from the average. You might like to sort the table entries into ascending order by grade before producing the output.

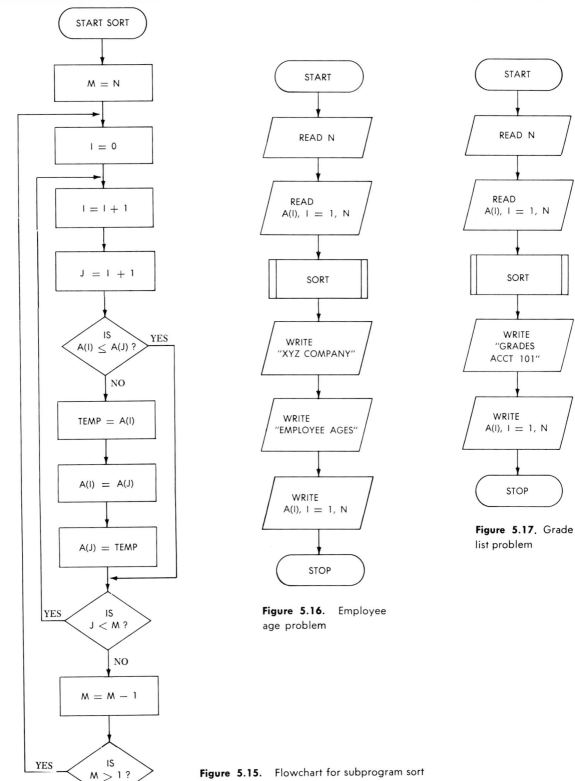

Figure 5.15. Flowchart for subprogram sort

Figure 5.16. Employee age problem

Figure 5.17. Grade list problem

2. A company has ten delivery trucks. It uses a 2 dimensional coordinate scheme to keep track of the location of each truck.

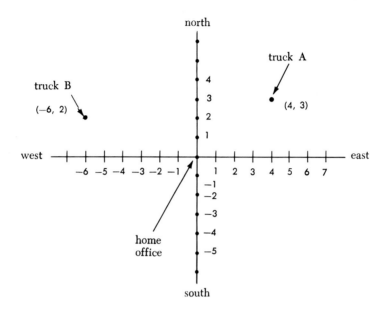

Draw a flowchart for a subroutine which could be used to calculate which truck is closest to the home office.

3. A truck has broken down. Because of the limited capacity of the trucks, the company needs to send two trucks to pick up the cargo of the disabled truck. Draw a flowchart for a subroutine to find the location of the truck closest to the broken truck. By eliminating the closest truck, a main program could call the subprogram a second time to find the location of the next closest truck.

APPENDIX A   Flowcharts and Programs

The representation of an algorithm in flowchart form is a convenient first step in solving problems in a computer environment. The second step consists of translating the flowchart into the language of the computer. Executable programs must ultimately be in *machine* language form. A machine language program is usually coded in terms of numeric operation codes and addresses. Writing a program in machine language is a difficult and time consuming task requiring extensive knowledge of the particular computer in use. Other languages that are problem oriented rather than machine oriented have been developed to allow the programmer to express the algorithm in a form easier to understand. Examples of widely used problem oriented languages are, business oriented languages such as COBOL (*COm-mon Business Oriented Language*), and algebraic languages such as FORTRAN (*FORmula TRANslation*) and BASIC (*Beginner's All-purpose Symbolic Instruction Code*).

A program written in any of these problem oriented languages must ultimately be translated into machine language before it can be processed by the CPU. Special programs (called *compilers*) have been devised to automatically translate programs written in languages such as COBOL, FORTRAN, and BASIC into machine language.

The algorithm of Figure 5.9 in chapter 5 was written in each of these three languages, and the resulting programs were executed with the sample data shown in Figure 5.13. The program listings and the output produced with the sample data is shown in Figures A.1, A.2, and A.3. Although the form of the programs differ significantly, the basic algorithm being implemented is the same, and essentially the same output is produced.

```
000010 IDENTIFICATION DIVISION.
000020 PROGRAM-ID. POSTAL.
000025 AUTHOR. GARY GLEASON.
000030 ENVIRONMENT DIVISION.
000070 INPUT-OUTPUT SECTION.
000080 FILE-CONTROL.
000100 SELECT CARD-FILE ASSIGN TO TERMINAL.
000110 SELECT PRINT-FILE ASSIGN TO PRINTER.
000130 DATA DIVISION.
000140 FILE SECTION.
000150 FD  PRINT-FILE
000160     LABEL RECORDS ARE OMITTED
000170     DATA RECORD IS PRINT-LINE.
000180 01  PRINT-LINE      PIC X(133).
000190 FD  CARD-FILE
000200     LABEL RECORDS ARE OMITTED
000210     DATA RECORDS ARE TABLE-VALUES, TRANSACTIONS.
000220 01  TABLE-VALUES.
000230     02 AMOUNT OCCURS 8 TIMES   PIC 9V99.
000240 01  TRANSACTIONS.
000250     02 Z           PIC 9.
000260     02 WEIGHT      PIC 9(3).
000270 WORKING-STORAGE SECTION.
000280 77 EOF-SWITCH      PIC 9 VALUE ZERO.
000290 77 SUM-IT          PIC 9(4)V99.
000300 77 TOTAL-COST      PIC 999V99 VALUE ZERO.
000310 01 TABLE-OF-COSTS.
000320     02 COST OCCURS 8 TIMES PIC 9V99.
000330 01 DETAIL-LINE.
000340     02 FILLER        PIC X(46) VALUE SPACES.
000350     02 Z-OUT         PIC 9.
000360     02 FILLER        PIC X(13) VALUE SPACES.
000370     02 WEIGHT-OUT    PIC ZZ9.
000380     02 FILLER        PIC X(13) VALUE SPACES.
000390     02 TOTAL-COST-OUT PIC $ZZZ.99.
000400 01 HEADING-1.
000410     02 FILLER        PIC X(46) VALUE SPACES.
000420     02 FILLER        PIC X(11) VALUE "XYZ COMPANY".
000430 01 HEADING-2.
000440     02 FILLER        PIC X(46) VALUE SPACES.
000450     02 FILLER        PIC X(4)  VALUE "ZONES".
000460     02 FILLER        PIC X(10) VALUE SPACES.
000470     02 FILLER        PIC X(6)  VALUE "WEIGHT".
000480     02 FILLER        PIC X(10) VALUE SPACES.
000490     02 FILLER        PIC X(10) VALUE "TOTAL COST".
000500 PROCEDURE DIVISION.
000510 MAJOR-LOGIC.
000520     OPEN OUTPUT PRINT-FILE.
000530     MOVE HEADING-1 TO PRINT-LINE.
000540     WRITE PRINT-LINE AFTER 1.
000550     MOVE HEADING-2 TO PRINT-LINE.
000560     WRITE PRINT-LINE AFTER 3.
000580     OPEN INPUT CARD-FILE.
000590     READ CARD-FILE AT END MOVE 1 TO EOF-SWITCH.
000600     MOVE TABLE-VALUES TO TABLE-OF-COSTS.
000610     READ CARD-FILE AT END MOVE 1 TO EOF-SWITCH.
000620     PERFORM CALCULATE-RATE UNTIL EOF-SWITCH = 1.
000630     CLOSE CARD-FILE.
000640     CLOSE PRINT-FILE.
000650     STOP RUN.
000660 CALCULATE-RATE.
000670     MULTIPLY WEIGHT BY COST (Z) GIVING TOTAL-COST.
000680     MOVE Z           TO Z-OUT.
000690     MOVE WEIGHT      TO WEIGHT-OUT.
000700     MOVE TOTAL-COST  TO TOTAL-COST-OUT.
000710     MOVE DETAIL-LINE TO PRINT-LINE.
000720     WRITE PRINT-LINE AFTER 2.
000730     READ CARD-FILE AT END MOVE 1 TO EOF-SWITCH.
```

XYZ COMPANY		
ZONE	WEIGHT	TOTAL COST
3	4	$ .56
7	16	$ 4.32

Figure A.1. STRUCTURED COBOL Program

```
 5    DIM C(8)
10    PRINT TAB(27); "XYZ COMPANY"
12    PRINT
15    PRINT TAB(21); "ZONE";TAB(27);"WEIGHT";TAB(36);"TOTAL COST"
20    FOR K = 1 TO 8
25    READ C(K)
30    NEXT K
35    READ Z, W
40    IF Z < 0 GO TO 65
45    LET T = W * C(Z)
50    PRINT TAB(22);Z;TAB(29);W;TAB(38);"$";T
55    GO TO 35
60    DATA .10, .12, .14, .17, .20, .24, .27, .35, 3, 4, 7, 16, -1, 1
65    END
RUN
```

                          XYZ COMPANY

ZONE	WEIGHT	TOTAL COST
3	4	$ .56
7	16	$ 4.32

Figure A.2.   BASIC program

```
      DIMENSION COST (8)
      INTEGER Z
      WRITE (3,2)
 2    FORMAT ('1', T61, 'XYZ COMPANY')
      WRITE (3,4)
 4    FORMAT (T47, 'ZONE', T61, 'WEIGHT', T76, 'TOTAL COST')
      READ (1,6) (COST (K), K = 1,8)
 6    FORMAT (8F2.2)
 8    READ (1, 10, END = 14) Z, WEIGHT
10    FORMAT (I1, F3.0)
      TOTCOS = WEIGHT * COST(Z)
      WRITE (3, 12) Z, WEIGHT, TOTCOS
12    FORMAT (T47, I1, T61, F4.0, T76, '$', F5.2)
      GO TO 8
14    STOP
      END
```

                          XYZ COMPANY

ZONE	WEIGHT	TOTAL COST
3	4	$ .56
7	16	$ 4.32

Figure A.3.   FORTRAN program

APPENDIX B   Introduction to BASIC

A BASIC program is a sequence of BASIC statements. Each statement must be preceded by a *sequence* number. Sequence numbers are sometimes referred to as *line* numbers or *statement* numbers. Sequence numbers are used to keep statements in proper logical sequence and to refer to one statement from another. BASIC statements may be entered in any desired order; they will be rearranged by the computer system into sequential order based on the sequence number. For example, the program:

```
10 INPUT X,Y
20 PRINT X,Y
30 END
```

could be entered as

```
20 PRINT X,Y
10 INPUT X,Y
30 END
```

or in any other desired sequence. When the system lists and executes the program, the statements will be in proper sequence.

B-1 VARIABLES

For purposes of flowcharting one may use any desired sequence of characters for variable names. In BASIC variable names must be either an alphabetic character (such as A, X, B, etc.) or an alphabetic character followed by a numeric character (such as A1, X2, Z9, etc.). Variable names such as PAY, HOURS, PERCENT, and so forth are invalid in BASIC. The programmer must substitute valid BASIC variable names when constructing programs from flowcharts having such descriptive variable names.

B-2 PRINT

The PRINT statement is used to produce output; it is the BASIC equivalent of /WRITE/ . Following are some examples of the PRINT statement:

20 PRINT X, Y, X+Y

The values of X and Y will be listed on the output device. The computation X+Y will be performed and the result will be printed.

40 PRINT "THE CENTIGRADE TEMPERATURE IS"; C

The output produced will be:
THE CENTIGRADE TEMPERATURE IS 20 ←value
contained in C

In BASIC the output line is divided into zones approximately 15 spaces in width. When a comma is used to separate elements in the PRINT list, the values will be printed in adjacent zones. When a semi-colon is used, elements are placed in the next available space.

B-3 INPUT

The INPUT statement is the BASIC equivalent to the flowcharting block /READ/ . Following are some examples of INPUT statements:

20 INPUT F

One value is entered by the user and stored by the computer in the variable F.

60 INPUT X,Y

Two values must be entered, the first will be stored in X and the second in Y.

When a terminal is used to execute a BASIC program, the system will type a *prompt* (usually a question mark) to alert the user to enter data. For example, consider the following statements:

10 PRINT "WHAT IS YOUR AGE";
20 INPUT A

When the above statements are executed, the results will be:

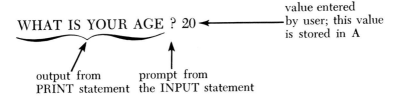

B-4 LET

Processing of data is performed by the LET statement which is used in the same way as a replacement statement in a flowchart. The general form of a LET statement is:

sequence-number LET variable = expression

In writing expressions, the following symbols are used to express arithmetic operations:

+   addition

—   subtraction

*   multiplication

/   division

↑   exponentiation

Following are some examples of valid LET statements:

30 LET P = H * R

This statement could be used to calculate the pay for an employee who worked H hours at rate of pay R.

40 LET Y = X ↑ 2

This statement is equivalent to $y = x^2$

In most BASIC systems the statement LET is optional, i.e., the statement 20 X = 1 is valid.

Parentheses may be used to indicate the order in which operations are performed. In any expression which does not contain parentheses, exponentiations will be performed first, then multiplications and divisions, and finally additions and subtractions. For example, the expression:

X * Y ↑ 2 − Z

will be evaluated as though parentheses had been inserted as follows:

(X * (Y ↑ 2)) − Z

As many sets of parentheses as are desired may be used. Parentheses may be necessary to force operations to be performed in the proper order. For example, in the statement

30 LET C = 5/9 * (F − 32)

the parentheses are required since 5/9 is to be multiplied by the value of F − 32. If parentheses had been omitted, 5/9 would be multiplied by F and then 32 would be subtracted which is not correct for computing the centigrade equivalent for a fahrenheit temperature.

B-5 BRANCHING

Unconditional branching is performed by the GO TO statement. The following are examples of valid GO TO statements:

30 GO TO 70

90 GO TO 20

→target statement

When a GO TO statement is executed, the program continues execution at the target statement.

Conditional branching is performed using the IF statement which is the BASIC equivalent of the flowcharting decision block ◇
The general form of the IF statement is:

$$\text{sequence-number IF condition} \begin{Bmatrix} \text{THEN} \\ \text{GO TO} \end{Bmatrix} \text{statement-number}$$

If the condition is true, the program branches to the specified statement; if the condition is false, the statement following the IF is executed next. The following symbols are used in the IF statement:

$>$	greater than
$<$	less than
$=$	equal to
$<\ >$	not equal to
$>\ =$	greater than or equal to
$<\ =$	less than or equal to

Following are some examples of IF statements:

    10 IF C = 1 THEN 120
    20 PRINT C

> If the value of C is 1 statement 120 is executed next, otherwise the value of C is printed.

    30 IF A + B < = C ↑ 2 GO TO 90
    40 IF A > C THEN 200

> If $A + B \leq C^2$ then statement 90 is executed next, otherwise the value of A is compared to the value of C.

B-6 TERMINATION

The END statement should be used as the last statement in every BASIC program. The END statement causes the execution of the program to cease. The STOP statement may be used anywhere in a program to cause the program to stop executing. As many STOP statements as are needed may be used in a program, however only one END statement may be used; in other respects the STOP and END statements have essentially the same function—to terminate the execution of a program.

B-7 SAMPLE PROGRAMS

The following program calculates an employee's pay and pays time and one half for hours over 40.

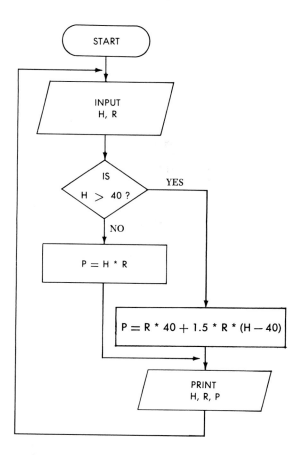

```
LIST    system command
100     INPUT H, R
200     IF H > 40 THEN 500
300     LET P = H * R
400     GO TO 600
500     LET P = R * 40 + 1.5 * R * (H — 40)
600     PRINT H, R, P
700     GO TO 100
800     END

RUN    system command
?34,6.0
 34            6            204
?52,4.00
 52            4            232
?
```

The commands LIST and RUN used in the above example are instructions to the computer system. LIST is used to obtain a listing of the program. The command RUN causes the program to be executed. Unlike BASIC statements no system commands should be numbered. The above program is caught in an infinite loop because of the GO TO statement at line 700. When calculations for one set of values has been completed the program asks for another set. When the user has no more data to be processed he enters a system command to cause the program to terminate. This system command is not shown in the above example because it is dependent on the particular computing system in use. Many systems use the Break key to stop an infinite loop. The reader may check with user's manuals written specifically for his system for more details.

The following program calculates the percentage of grades which are greater than 75. A negative value is used to signify the end of the data file.

```
LIST
100    LET  T=0
200    LET  K=0
300    INPUT  G
400    IF G < 0 THEN 900
500    LET T = T + 1
600    IF G <= 75 THEN 300
700    LET K=K + 1
800    GO TO 300
900    LET P = K / T * 100
1000    PRINT "PERCENTAGE = "; P
1100    END

RUN
?45
?89
?35
?75
?100
?—1
PERCENTAGE = 40
```

B-8 FOR/NEXT

The FOR and NEXT statements are used in BASIC for automatic loop control. The FOR statement specifies a variable name, the initial value for that variable, the terminal value, and the increment. For example, the statement:

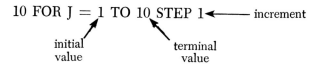

10 FOR J = 1 TO 10 STEP 1 ←——— increment

       initial   terminal
       value   value

is equivalent to the flowchart block:

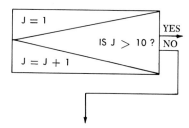

The STEP portion of the FOR statement is optional; if omitted, the increment is assumed to be 1. Thus the statement:

10 FOR J = 1 TO 10

is equivalent to the above FOR statement.

The NEXT statement marks the end of the body of statements to be repeated. For example, the following program will print the numbers from 1 to 10:

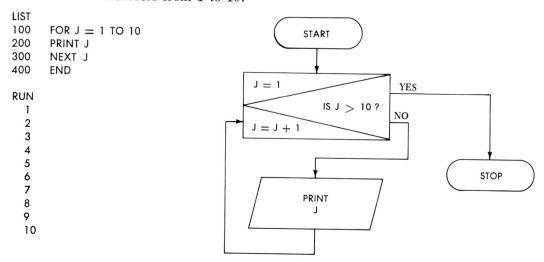

```
LIST
100    FOR J = 1 TO 10
200    PRINT J
300    NEXT J
400    END

RUN
 1
 2
 3
 4
 5
 6
 7
 8
 9
10
```

The program shown below produces a table showing the correspondence between Fahrenheit and centigrade temperatures for values ranging from 0° F to 5° F in increments of .5°.

```
LIST
100    FOR F = 0 TO 5 STEP . 5
200    LET C = 5 / 9 * (F — 32)
300    PRINT  F, C
400    NEXT  F
500    END

RUN
0                —17.777777778
0.5              —17.5
1                —17.222222222
1.5              —16.944444444
2                —16.666666667
2.5              —16.388888889
3                —16.111111111
3.5              —15.833333333
4                —15.555555556
4.5              —15.277777778
5                —15
```

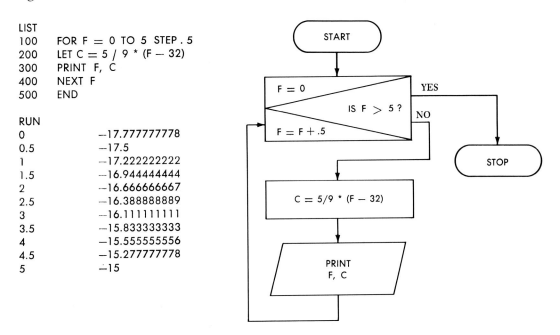

The following program is capable of producing tables of Fahrenheit and centigrade temperatures with the starting value N, the terminating value M, and the increment I entered by the program user. Note the FOR statement at line 200 which controls the value of the variable F at line 300; as is illustrated in this statement, the initial, terminal, and increment values may be specified by variables as well as constants.

```
LIST
100     INPUT N, M, I
200     FOR F = N TO M STEP I
300     LET C = 5/9 * (F — 32)
400     PRINT F, C
500     NEXT F
600     GO TO 100
700     END

RUN
?0,250,25
 0                      —17.777777778
 25                     —3.8888888889
 50                      10
 75                      23.888888889
 100                     37.777777778
 125                     51.666666667
 150                     65.555555555
 175                     79.444444444
 200                     93.333333332
 225                     107.22222222
 250                     121.11111111
?70,110,5
 70                      21.111111111
 75                      23.888888889
 80                      26.666666667
 85                      29.444444444
 90                      32.222222222
 95                      35
 100                     37.777777778
 105                     40.555555556
 110                     43.333333333
```

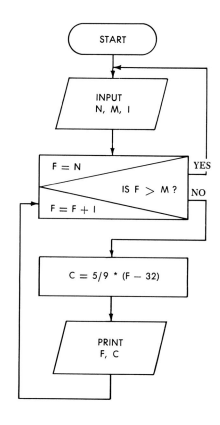

APPENDIX C   Decision Tables

A decision table is a formal technique used to describe rules which govern actions to be taken in a given context. Decision tables are often used with or in lieu of verbal descriptions and flowcharts. In many instances a decision table can be a useful intermediate form between a verbal description of a procedure and a flowchart for a program to implement the procedure.

A decision table shows a set of conditions, a set of actions, and rules which link the conditions to the actions. For example, consider the familiar procedure for pay calculation:

> If the number of hours worked in a week is over 40, the employee is paid time and one-half for hours in excess of 40, otherwise he is paid at regular rate for hours worked.

In this situation there are two conditions:

> hours worked less than or equal to 40
> hours worked greater than 40

and two actions:

> pay at regular rate
> pay at regular rate for 40 hours and 1½ times regular rate for hours over 40.

The procedure can be expressed in tabular form as shown in Figure C.1. The interpretation of the columns labeled "Rules" is as follows:
If these conditions are satisfied, take the indicated actions.
Thus Rule 1 is:

> If hours ≤ 40 calculate pay at regular rate.

Payroll Calculations		Rules	
		1	2
Conditions	Hours ≤ 40	x	
	Hours > 40		x
Actions	Regular rate	x	
	Regular rate for 40 hours plus 1½ times regular rate for hours over 40		x

Figure C.1. Decision table for payroll calculation

Leap Year Calculation		Rules			
		1	2	3	4
Conditions	Year divisible by 4		x		
	Year not divisible by 4	x			
	Year divisible by 100				x
	Year not divisible by 100		x		
	Year divisible by 400			x	
	Year not divisible by 400				x
Actions	Leap year		x	x	
	Not leap year	x			x

Figure C.2. Decision table for leap year calculation

Rule 2 is:

> If hours > 40 calculate pay at regular rate for 40 hours plus 1½ times regular rate for hours over 40.

The simplicity of the above example tends to hide the power and usefulness of the decision table technique. Consider the slightly more complex procedure for determining whether or not a year is a leap year:

> If a year is divisible by 4 it is a leap year, with the following exceptions: years that are divisible by 100 but not 400 are not leap years.

Thus the year 1900 was not a leap year but the year 2000 (which is divisible by 400) will be a leap year. Confused? Perhaps a decision table will help in making the procedure clearer. In this case there are six conditions and two actions as shown in the decision table in Figure C.2. In order to determine whether or not a year is a leap year it is necessary to find out which of the four rules applies as shown in the examples below.

> Examples
>
> 1981 fits the conditions of Rule 1 (not divisible by 4) and hence is not a leap year.
>
> 1984 complies with the conditions of Rule 2 (divisible by 4 but not by 100) and hence is a leap year.
>
> 2000 satisfies the conditions expressed for Rule 3 (divisible by 400) and hence is a leap year.
>
> 1900 fits the conditions of Rule 4 (divisible by 100 but not 400) and is not a leap year.

In many instances it is easier to convert decision tables into flowchart form than a verbal description of the same procedure. For example, the flowchart form of the leap year calculation is shown in Figure C.3 and is a straightforward translation of the four rules into a sequence of decisions.

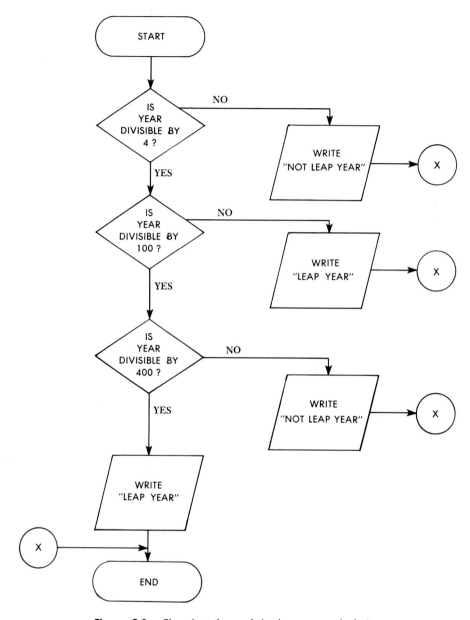

Figure C.3. Flowchart form of the leap year calculation

In both examples considered so far there have only been two alternative actions which were selected by the rules specified. In general, there may be any number of alternative actions or combinations of actions to be selected. For example, consider the following policy for determination of Christmas bonuses for employees of XYZ Manufacturing Corporation:

> To qualify for any bonus the employee must have been employed at least six months. A bonus of one week's pay will be paid to hourly employees with a satisfactory rating; hourly employees with an unsatisfactory rating will receive one day's pay. Salaried employees receive 20% of monthly pay if their rating is satisfactory but only 5% of monthly pay if their rating is unsatisfactory. Employees with more than ten years of service and a satisfactory rating receive an additional $400 bonus.

The decision table for this procedure is shown in Figure C.4. Note Rules 4 and 7 which specify that two actions are to be taken when this set of conditions is found.

In general, decision tables are a useful intermediate step in the analysis of a problem because they force the programmer to explicitly state all of the relevant conditions and the rules which govern the required outcomes. Constructing a decision table for a procedure will very often help to ensure that the flowchart and the program written are logically correct and implement the specified procedure exactly. A decision table can also help in determining inconsistencies (when two rules conflict with each other) and incompleteness (when no action is specified for a set of conditions) in the original statement of the procedure. For an example of an incomplete problem statement, construct a decision table for the metal grading problem (Problem 14 Exercises 2-2) and note that some combinations of conditions have no action specified for them.

Christmas Bonus Calculation		Rules						
		1	2	3	4	5	6	7
Conditions	Employment < 6 mo.	x						
	Employment > 10 yr.				x			x
	6 mo. ≤ Employment ≤ 10 yr.		x	x		x	x	
	Hourly employee		x	x	x			
	Salaried employee					x	x	x
	Rating is satisfactory		x		x	x		x
	Rating is unsatisfactory			x			x	
Actions	No bonus	x						
	One week's pay		x		x			
	One day's pay			x				
	20% of monthly pay					x		x
	5% of monthly pay						x	
	$400 bonus				x			x

Figure C.4. Decision table for Christmas bonus calculation

Solutions to Selected Exercises

EXERCISES 1-1

1. 170
3. 1. READ BASE,ALT
   2. AREA=BASE*ALT/2
   3. WRITE AREA
   4. STOP

   Note: The "*" is often used to denote multiplication;
   the "/" is used to denote division. See section 2-3.

EXERCISES 2-1

3. a. Invalid. There is a constant on the left side of the replacement statement.
   c. Invalid. There is an expression on the left side of the replacement statement.
   e. Valid.
   f. Invalid. The expression is not a replacement statement.

6. In Figure 2.3 the value of X is assigned by a READ statement; in Figure 2.12 a replacement statement is used. Figure 2.12 does not represent an algorithm because it lacks input.

9. a. 400, 40, 10
   b. 3, 4, 25
   c. 73, 31, 42

11. a. Invalid. "4" is a constant, not a variable.
    b. Invalid. "X=4" is a replacement statement not a variable.
    c. Valid.
    d. Valid.

12. a. A=12, B=11.6, C= −13
    b. HRS=12, RATE=11.6
    c. A=3, B=0 Note: When a second value is read for a variable it replaces the value previously stored in that variable.
    d. GRADE= − 3

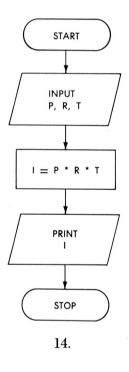

14.

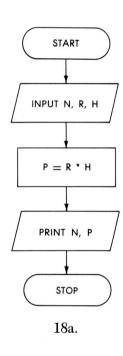

18a.

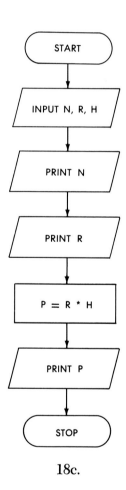

18c.

EXERCISES 2-2

7.

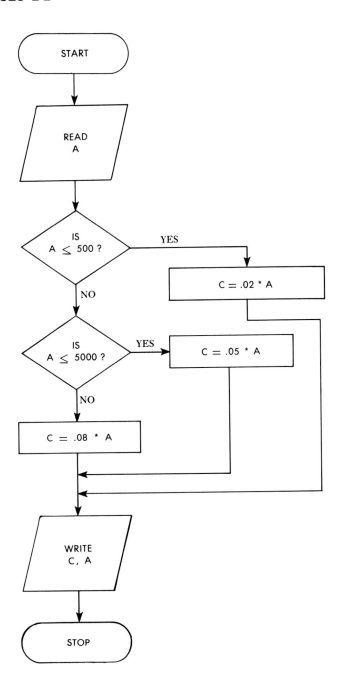

10.

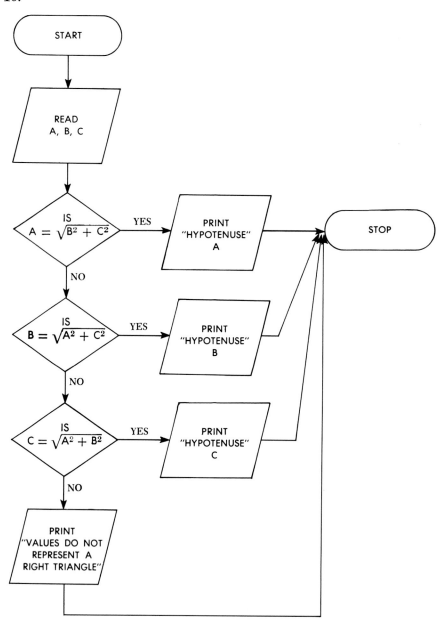

11.

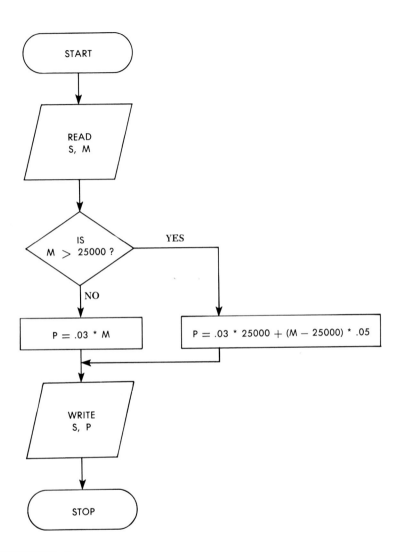

EXERCISES 3-1

3. No, because the statement I = 0 is in the body of the loop the value of I will be reinitialized to zero with each repetition of the loop.

5. It is not necessary to initialize a counter to 1 or 0. The counter can be initialized to some larger value and then decremented each time the loop is processed. A test is made to determine when the counter reaches 1 to halt repetition of the loop. This problem loops five times.

15.

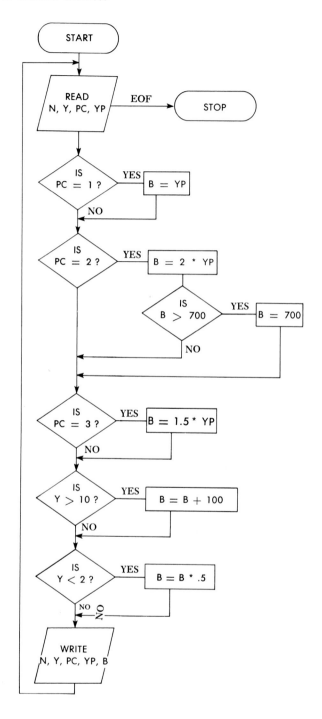

EXERCISES 3-2

4a.

4d.

4h.

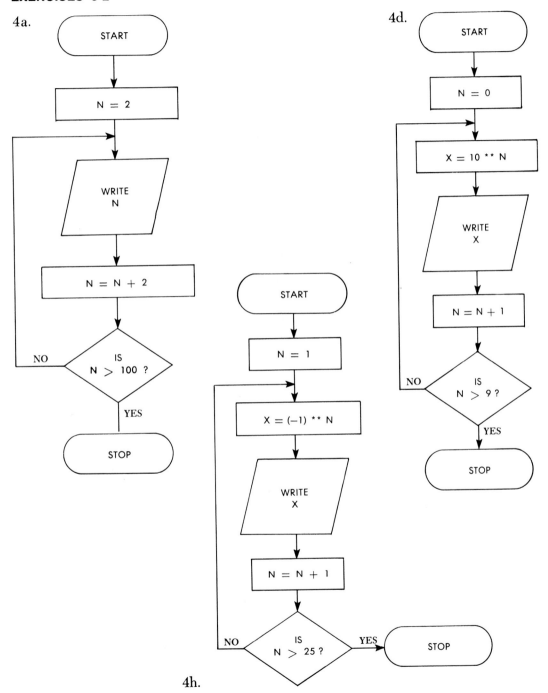

10a.

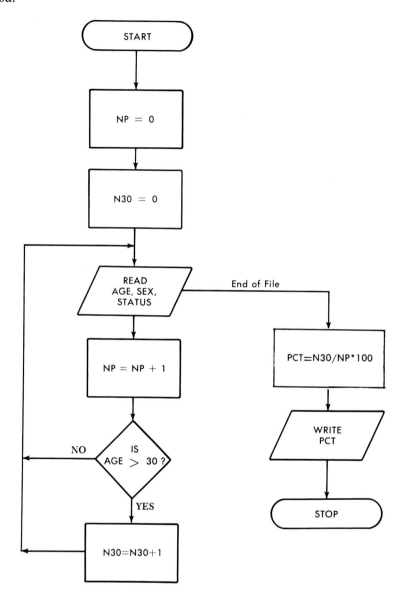

10c.

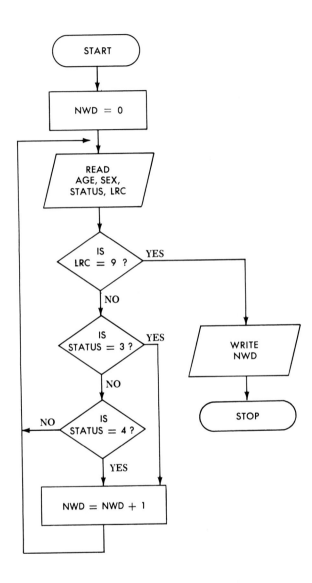

10d.

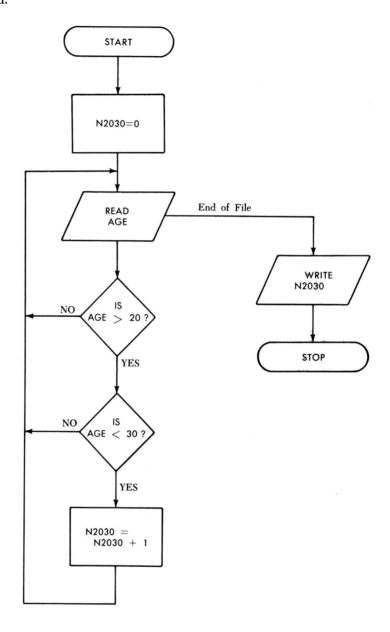

17.

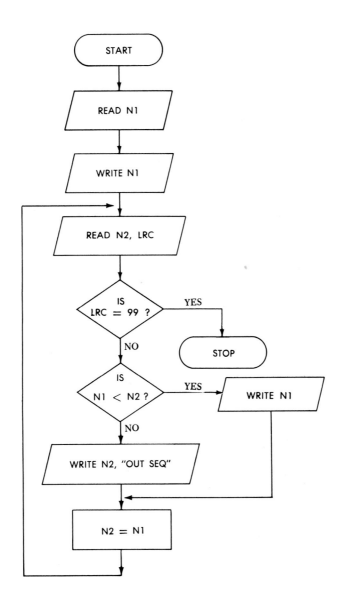

EXERCISES 3-3

3a.

13.

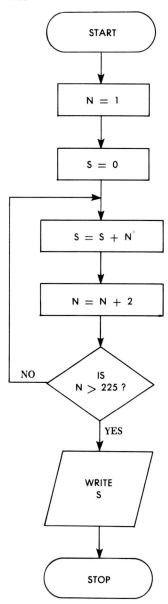

EXERCISES 3-4

2.

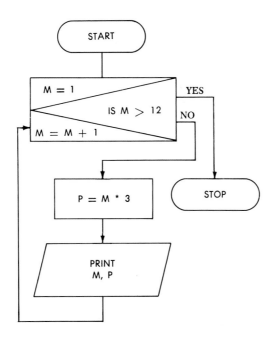

EXERCISES 4-1

1. System Flowchart:

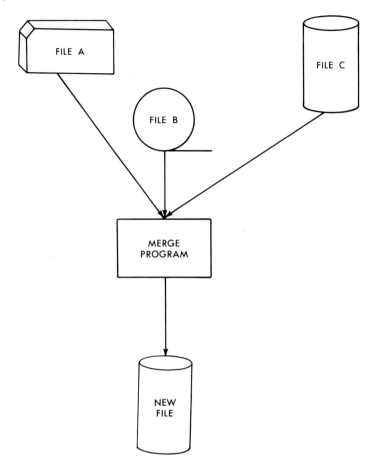

1. Program Flowchart:

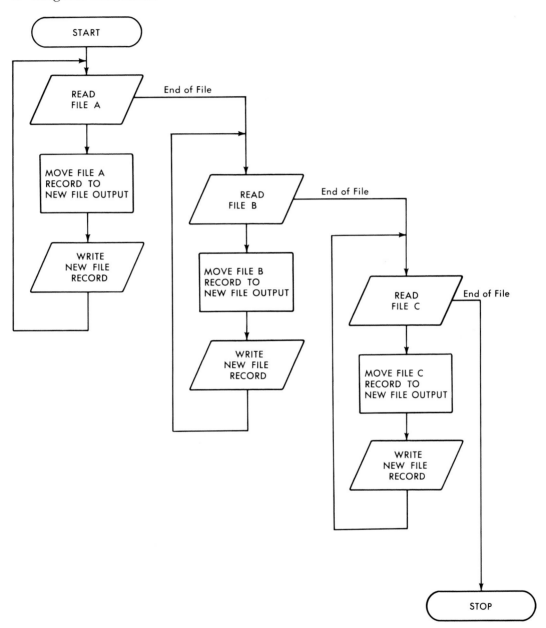

EXERCISES 4-2

4.

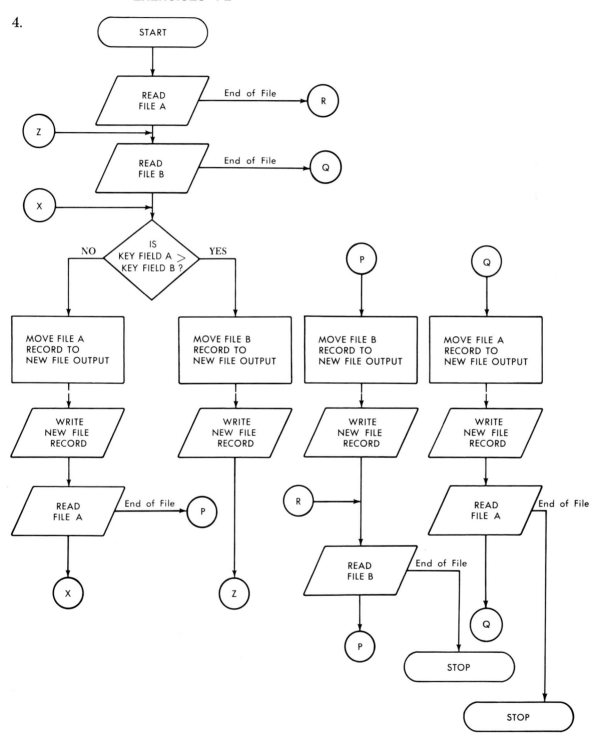

5.

	ACCT12		ACCT19		ACCT20		ACCT30	

ACCT40		ACCT50		ACCT51		/*	

7.

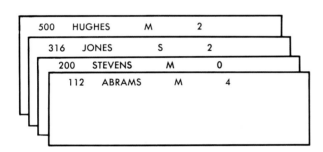

```
500    HUGHES      M         2
   316    JONES        S         2
      200    STEVENS    M         0
         112    ABRAMS      M         4
```

EXERCISES 5-1

3. Method 1

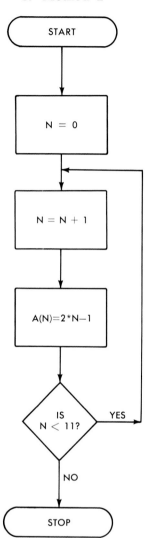

Method 2

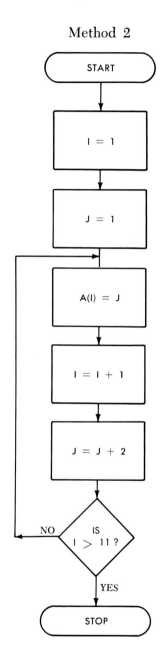

7.

Z	WEIGHT	TOTCOST
3	4	.56
7	16	4.32

9.

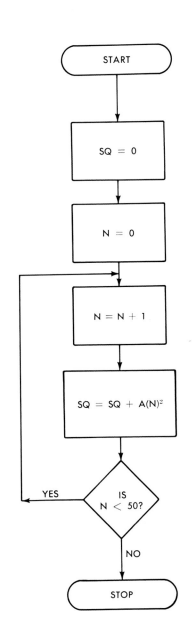

11. Method 1

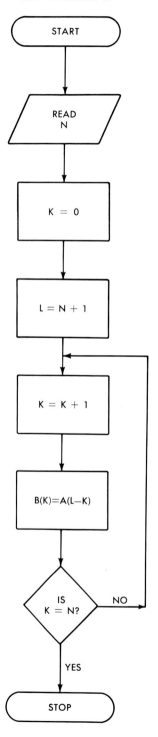

Method 2

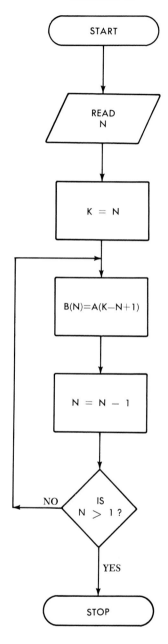

Index